DESIGN AND ANALYSIS OF EXPERIMENTS

8th Edition

STUDENT SOLUTIONS MANUAL

Lisa Custer

Accenture

Daniel R. McCarville

Arizona State University

Douglas C. Montgomery

Arizona State University

WILEY

Founded in 1807, John Wiley & Sons, Inc. has been a valued source of knowledge and understanding for more than 200 years, helping people around the world meet their needs and fulfill their aspirations. Our company is built on a foundation of principles that include responsibility to the communities we serve and where we live and work. In 2008, we launched a Corporate Citizenship Initiative, a global effort to address the environmental, social, economic, and ethical challenges we face in our business. Among the issues we are addressing are carbon impact, paper specifications and procurement, ethical conduct within our business and among our vendors, and community and charitable support. For more information, please visit our website: www.wiley.com/go/citizenship.

ISBN-978-1-118-38819-8

10 9 8 7 6 5 4 3 2 1

CHAPTER **1**

Introduction

LEARNING OBJECTIVES

After completing this chapter, you will be able to:

1. Understand the strategy of experimentation and know how factorial experimental designs compare to best-guess and one-factor-at-a-time approaches.

2. Describe the three basic principles of designed experiments.

3. Understand how the seven-step process for planning, conducting, and analyzing the data from a designed experiment is used.

4. Understand the difference between the agricultural heritage of statistically designed experiments and how the techniques are used in the modern industrial environment.

5. Appreciate how statistical techniques are used in industrial experimentation.

KEY CONCEPTS AND IDEAS

1. Best-guess experiment
2. One-factor-at-a-time experiments
3. Factorial experiments
4. Interaction
5. Fractional factorial experiments
6. Screening experiments
7. Optimization experiments
8. Response surface
9. Replication
10. Randomization
11. Blocking
12. Sequential experimentation

CHAPTER 2

Simple Comparative Experiments

LEARNING OBJECTIVES

After completing this chapter, you will be able to:

1. Use basic statistical concepts including random sampling, probability distributions, random samples, sampling distributions, and tests of hypotheses.

2. Use the two-sample or pooled t-test to compare two means when the experiment is conducted in a completely randomized design.

3. Use the paired t-test to compare the difference in two means when the experiment is conducted in a paired comparison design.

4. Use confidence intervals to express the difference in means.

5. Check the assumptions for the t-test.

KEY CONCEPTS AND IDEAS

1. Treatment
2. Factor level
3. Statistical hypothesis testing
4. Confidence interval
5. P-value
6. Normal distribution
7. Chi-square distribution
8. t-distribution
9. F-distribution
10. Completely randomized experiments
11. Paired comparison designs
12. Normal probability plot
13. Statistical model

Exercises

2.20. The shelf life of a carbonated beverage is of interest. Ten bottles are randomly selected and tested, and the following results are obtained:

Days	
108	138
124	163
124	159
106	134
115	139

(a) We would like to demonstrate that the mean shelf life exceeds 120 days. Set up appropriate hypotheses for investigating this claim.

$$H_0: \mu = 120 \qquad H_1: \mu > 120$$

(b) Test these hypotheses using $\alpha = 0.01$. What are your conclusions?

$\bar{y} = 131$

$S^2 = [(108 - 131)^2 + (124 - 131)^2 + (124 - 131)^2 + (106 - 131)^2 + (115 - 131)^2 + (138 - 131)^2$
$\qquad + (163 - 131)^2 + (159 - 131)^2 + (134 - 131)^2 + (139 - 131)^2] / (10 - 1)$

$S^2 = 3438 / 9 = 382$

$S = \sqrt{382} = 19.54$

$$t_0 = \frac{\bar{y} - \mu_0}{S/\sqrt{n}} = \frac{131 - 120}{19.54/\sqrt{10}} = 1.78$$

Because $t_{0.01,9} = 2.821$, there is no strong evidence to support the conclusion that mean shelf life exceeds 120 days. Do not reject H_0.

MINITAB Output

T-Test of the Mean

Test of mu = 120.00 vs mu > 120.00

Variable	N	Mean	StDev	SE Mean	T	P
Shelf Life	10	131.00	19.54	6.18	1.78	0.054

T Confidence Intervals

Variable	N	Mean	StDev	SE Mean	99.0 % CI	
Shelf Life	10	131.00	19.54	6.18	(110.91,	151.09)

(c) Find the *P*-value for the test in part (b).

$P = 0.054$

(d) Construct a 99 percent confidence interval on the mean shelf life.

The 99% confidence interval is $\bar{y} - t_{\alpha/2, n-1} \dfrac{S}{\sqrt{n}} \le \mu \le \bar{y} + t_{\alpha/2, n-1} \dfrac{S}{\sqrt{n}}$ with $\alpha = 0.01$.

$$131 - (3.250)\left(\frac{19.54}{\sqrt{10}}\right) \le \mu \le 131 + (3.250)\left(\frac{19.54}{\sqrt{10}}\right)$$

$$110.91 \le \mu \le 151.08$$

2.26. The following are the burning times (in minutes) of chemical flares of two different formulations. The design engineers are interested in both the mean and variance of the burning times.

Type 1		Type 2	
65	82	64	56
81	67	71	69
57	59	83	74
66	75	59	82
82	70	65	79

(a) Test the hypotheses that the two variances are equal. Use $\alpha = 0.05$.

$$H_0 : \sigma_1^2 = \sigma_2^2 \qquad S_1 = 9.264$$
$$H_1 : \sigma_1^2 \ne \sigma_2^2 \qquad S_2 = 9.367$$
$$F_0 = \frac{S_1^2}{S_2^2} = \frac{85.82}{87.73} = 0.98$$

$$F_{0.025,9,9} = 4.03 \qquad F_{0.975,9,9} = \frac{1}{F_{0.025,9,9}} = \frac{1}{4.03} = 0.248$$

Do not reject H_0. There is no strong evidence to indicate that the two variances are different.

(b) Using the results of (a), test the hypotheses that the mean burning times are equal. Use $\alpha = 0.05$. What is the P-value for this test?

$$S_p^2 = \frac{(n_1 - 1)S_1^2 + (n_2 - 1)S_2^2}{n_1 + n_2 - 2} = \frac{1561.95}{18} = 86.775$$

$$S_p = 9.32$$

$$t_0 = \frac{\bar{y}_1 - \bar{y}_2}{S_p\sqrt{\dfrac{1}{n_1} + \dfrac{1}{n_2}}} = \frac{70.4 - 70.2}{9.32\sqrt{\dfrac{1}{10} + \dfrac{1}{10}}} = 0.048$$

$$t_{0.025,18} = 2.101 \quad \text{Do not reject.}$$

From the computer output, $t=0.05$; do not reject. Also from the MINITAB output below, $P=0.96$

MINITAB Output

Two Sample T-Test and Confidence Interval

```
Two sample T for Type 1 vs Type 2

          N       Mean      StDev    SE Mean
Type 1   10      70.40       9.26        2.9
Type 2   10      70.20       9.37        3.0

95% CI for mu Type 1 - mu Type 2: ( -8.6,   9.0)
T-Test mu Type 1 = mu Type 2 (vs not =): T = 0.05   P = 0.96   DF = 18
Both use Pooled StDev = 9.32
```

(c) Discuss the role of the normality assumption in this problem. Check the assumption of normality for both types of flares.

The assumption of normality is required in the theoretical development of the t-test. However, moderate departure from normality has little impact on the performance of the t-test. The normality assumption is more important for the test on the equality of the two variances. An indication of nonnormality would be of concern here. The normal probability plots shown below indicate that burning times for both formulations follow the normal distribution.

Normal Probability Plot

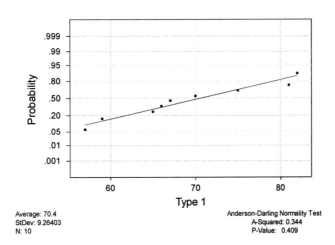

Average: 70.4
StDev: 9.26403
N: 10

Anderson-Darling Normality Test
A-Squared: 0.344
P-Value: 0.409

Normal Probability Plot

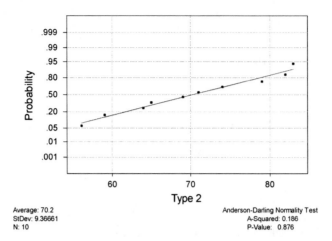

Average: 70.2
StDev: 9.36661
N: 10

Anderson-Darling Normality Test
A-Squared: 0.186
P-Value: 0.876

2.27. An article in *Solid State Technology*, "Orthogonal Design of Process Optimization and Its Application to Plasma Etching" by G.Z. Yin and D.W. Jillie (May, 1987) describes an experiment to determine the effect of C_2F_6 flow rate on the uniformity of the etch on a silicon wafer used in integrated circuit manufacturing. Data for two flow rates are as follows:

C_2F_6	Uniformity Observation					
(SCCM)	1	2	3	4	5	6
125	2.7	4.6	2.6	3.0	3.2	3.8
200	4.6	3.4	2.9	3.5	4.1	5.1

(a) Does the C_2F_6 flow rate affect average etch uniformity? Use $\alpha = 0.05$.

No, C_2F_6 flow rate does not affect average etch uniformity.

MINITAB Output

```
Two Sample T-Test and Confidence Interval

Two sample T for Uniformity

Flow Rat    N      Mean     StDev    SE Mean
125         6      3.317    0.760    0.31
200         6      3.933    0.821    0.34

95% CI for mu (125) - mu (200): ( -1.63,  0.40)
T-Test mu (125) = mu (200) (vs not =): T = -1.35   P = 0.21   DF = 10
Both use Pooled StDev = 0.791
```

(b) What is the *P*-value for the test in part (a)?

From the MINITAB output, *P*=0.21.

(c) Does the C_2F_6 flow rate affect the wafer-to-wafer variability in etch uniformity? Use $\alpha = 0.05$.

$$H_0 : \sigma_1^2 = \sigma_2^2$$
$$H_1 : \sigma_1^2 \neq \sigma_2^2$$
$$F_{0.025,5,5} = 7.15$$
$$F_{0.975,5,5} = 0.14$$
$$F_0 = \frac{0.5776}{0.6724} = 0.86$$

Do not reject; C_2F_6 flow rate does not affect wafer-to-wafer variability.

(d) Draw box plots to assist in the interpretation of the data from this experiment.

The box plots shown below indicate that there is little difference in uniformity at the two gas flow rates. Any observed difference is not statistically significant. See the t-test in part (a).

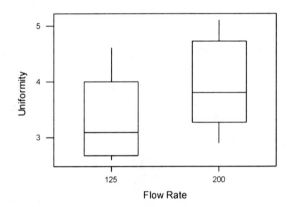

2.30. Front housings for cell phones are manufactured in an injection molding process. The time the part is allowed to cool in the mold before removal is thought to influence the occurrence of a particularly troublesome cosmetic defect, flow lines, in the finished housing. After manufacturing, the housings are inspected visually and assigned a score between 1 and 10 based on their appearance, with 10 corresponding to a perfect part and 1 corresponding to a completely defective part. An experiment was conducted using two cool-down times, 10 seconds and 20 seconds, and 20 housings were evaluated at each level of cool-down time. The data are shown below.

10 Seconds		20 Seconds	
1	3	7	6
2	6	8	9
1	5	5	5
3	3	9	7
5	2	5	4
1	1	8	6
5	6	6	8
2	8	4	5
3	2	6	8
5	3	7	7

(a) Is there evidence to support the claim that the longer cool-down time results in fewer appearance defects? Use $\alpha = 0.05$.

From the analysis shown below, there is evidence that the longer cool-down time results in fewer appearance defects.

MINITAB Output

```
Two-Sample T-Test and CI: 10 seconds, 20 seconds

Two-sample T for 10 seconds vs 20 seconds

            N     Mean    StDev   SE Mean
10 secon   20     3.35    2.01      0.45
20 secon   20     6.50    1.54      0.34


Difference = mu 10 seconds - mu 20 seconds
Estimate for difference:  -3.150
95% upper bound for difference: -2.196
T-Test of difference = 0 (vs <): T-Value = -5.57   P-Value = 0.000   DF = 38
Both use Pooled StDev = 1.79
```

(b) What is the P-value for the test conducted in part (a)?

From the MINITAB output, $P = 0.000$.

(c) Find a 95% confidence interval on the difference in means. Provide a practical interpretation of this interval.

From the MINITAB output, $\mu_1 - \mu_2 \leq -2.196$. This lower confidence bound is less than 0. The two samples are different. The 20 second cooling time gives a cosmetically better housing.

(d) Draw dot diagrams to assist in interpreting the results from this experiment.

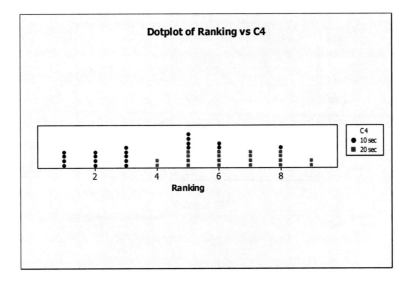

Dotplot of Ranking vs C4

(e) Check the assumption of normality for the data from this experiment.

Normal Probability Plot

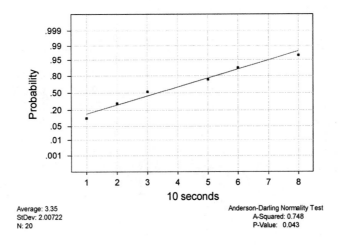

Average: 3.35
StDev: 2.00722
N: 20

Anderson-Darling Normality Test
A-Squared: 0.748
P-Value: 0.043

Normal Probability Plot

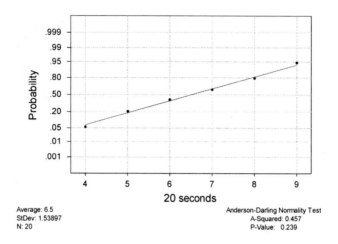

Average: 6.5
StDev: 1.53897
N: 20

Anderson-Darling Normality Test
A-Squared: 0.457
P-Value: 0.239

There are no significant departures from normality.

2.34. An article in the *Journal of Strain Analysis* (vol.18, no. 2, 1983) compares several procedures for predicting the shear strength for steel plate girders. Data for nine girders in the form of the ratio of predicted to observed load for two of these procedures, the Karlsruhe and Lehigh methods, are as follows:

Girder	Karlsruhe Method	Lehigh Method	Difference	Difference^2
S1/1	1.186	1.061	0.125	0.015625
S2/1	1.151	0.992	0.159	0.025281
S3/1	1.322	1.063	0.259	0.067081
S4/1	1.339	1.062	0.277	0.076729
S5/1	1.200	1.065	0.135	0.018225
S2/1	1.402	1.178	0.224	0.050176
S2/2	1.365	1.037	0.328	0.107584
S2/3	1.537	1.086	0.451	0.203401
S2/4	1.559	1.052	0.507	0.257049
		Sum =	2.465	0.821151
		Average =	0.274	

(a) Is there any evidence to support a claim that there is a difference in mean performance between the two methods? Use $\alpha = 0.05$.

$$H_0: \mu_1 = \mu_2 \quad \text{or equivalently} \quad H_0: \mu_d = 0$$
$$H_1: \mu_1 \neq \mu_2 \qquad\qquad\qquad H_1: \mu_d \neq 0$$

$$\bar{d} = \frac{1}{n}\sum_{i=1}^{n} d_i = \frac{1}{9}(2.465) = 0.274$$

$$s_d = \left[\frac{\sum\limits_{i=1}^{n} d_i^2 - \frac{1}{n}\left(\sum\limits_{i=1}^{n} d_i\right)^2}{n-1}\right]^{1/2} = \left[\frac{0.821151 - \frac{1}{9}(2.465)^2}{9-1}\right]^{1/2} = 0.135$$

$$t_0 = \frac{\bar{d}}{\frac{S_d}{\sqrt{n}}} = \frac{0.274}{\frac{0.135}{\sqrt{9}}} = 6.08$$

$$t_{\alpha/2, n-1} = t_{0.025, 8} = 2.306 \text{, reject the null hypothesis.}$$

MINITAB Output

Paired T-Test and Confidence Interval

Paired T for Karlsruhe - Lehigh

	N	Mean	StDev	SE Mean
Karlsruh	9	1.3401	0.1460	0.0487
Lehigh	9	1.0662	0.0494	0.0165
Difference	9	0.2739	0.1351	0.0450

95% CI for mean difference: (0.1700, 0.3777)
T-Test of mean difference = 0 (vs not = 0): T-Value = 6.08 P-Value = 0.000

(b) What is the P-value for the test in part (a)?

$P=0.0002$

(c) Construct a 95 percent confidence interval for the difference in mean predicted to observed load.

$$\bar{d} - t_{\alpha/2, n-1}\frac{S_d}{\sqrt{n}} \le \mu_d \le \bar{d} + t_{\alpha/2, n-1}\frac{S_d}{\sqrt{n}}$$

$$0.274 - 2.306\frac{0.135}{\sqrt{9}} \le \mu_d \le 0.274 + 2.306\frac{0.135}{\sqrt{9}}$$

$$0.17023 \le \mu_d \le 0.37777$$

(d) Investigate the normality assumption for both samples.

The normal probability plots of the observations for each method follows. There are no serious concerns with the normality assumption, but there is an indication of a possible outlier (1.178) in the Lehigh method data.

Normal Probability Plot

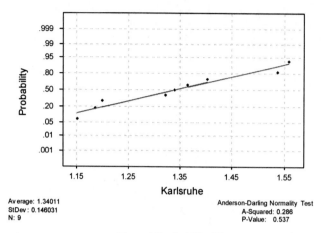

Karlsruhe

Average: 1.34011
StDev: 0.146031
N: 9

Anderson-Darling Normality Test
A-Squared: 0.286
P-Value: 0.537

Normal Probability Plot

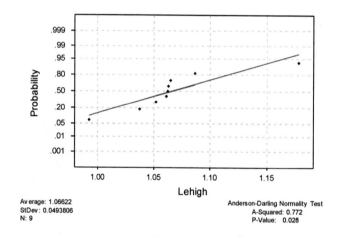

Lehigh

Average: 1.06622
StDev: 0.0493806
N: 9

Anderson-Darling Normality Test
A-Squared: 0.772
P-Value: 0.028

(e) Investigate the normality assumption for the difference in ratios for the two methods.

The normal probability plot follows. There is no indication of a problem with the normality assumption.

Normal Probability Plot

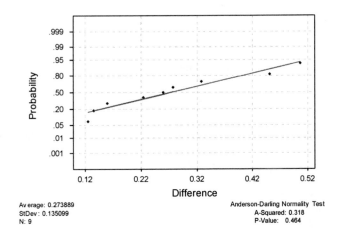

Average: 0.273889
StDev: 0.135099
N: 9

Anderson-Darling Normality Test
A-Squared: 0.318
P-Value: 0.464

(f) Discuss the role of the normality assumption in the paired t-test.

As in any t-test, the assumption of normality is only moderately important. In the paired t-test, the assumption of normality applies to the distribution of the differences. That is, the individual sample measurements do not have to be normally distributed, only their difference.

2.37. In semiconductor manufacturing, wet chemical etching is often used to remove silicon from the backs of wafers prior to metallization. The etch rate is an important characteristic of this process. Two different etching solutions are being evaluated. Eight randomly selected wafers have been etched in each solution and the observed etch rates (in mils/min) are shown as follows.

Solution 1		Solution 2	
9.9	10.6	10.2	10.6
9.4	10.3	10.0	10.2
10.0	9.3	10.7	10.4
10.3	9.8	10.5	10.3

(a) Do the data indicate that the claim that both solutions have the same mean etch rate is valid? Use $\alpha = 0.05$ and assume equal variances.

No, the solutions do not have the same mean etch rate. See the MINITAB output below.

MINITAB Output

```
Two Sample T-Test and Confidence Interval

Two-sample T for Solution 1 vs Solution 2

           N    Mean   StDev   SE Mean
Solution   8   9.950   0.450     0.16
Solution   8  10.363   0.233     0.082
```

```
Difference = mu Solution 1 - mu Solution 2
Estimate for difference:  -0.413
95% CI for difference: (-0.797, -0.028)
T-Test of difference = 0 (vs not =): T-Value = -2.30  P-Value = 0.037  DF = 14
Both use Pooled StDev = 0.358
```

(b) Find a 95% confidence interval on the difference in mean etch rate.

From the MINITAB output, -0.797 to –0.028.

(c) Use normal probability plots to investigate the adequacy of the assumptions of normality and equal variances.

Normal Probability Plot

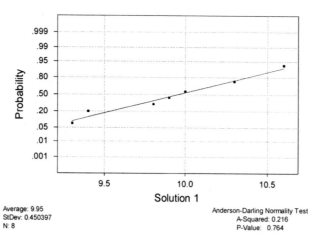

Average: 9.95
StDev: 0.450397
N: 8

Anderson-Darling Normality Test
A-Squared: 0.216
P-Value: 0.764

Normal Probability Plot

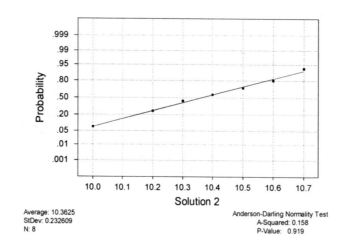

Average: 10.3625
StDev: 0.232609
N: 8

Anderson-Darling Normality Test
A-Squared: 0.158
P-Value: 0.919

Both the normality and equality of variance assumptions are valid.

2.46. Consider the experiment described in problem 2.26. If the mean burning times of the two flames differ by as much as 2 minutes, find the power of the test. What sample size would be required to detect an actual difference in mean burning time of 1 minute with a power of at least 0.90?

From the MINITAB output below, the power is 0.0740. This answer was obtained by using the pooled estimate of σ from Problem 2.21, $S_p = 9.32$. Because the difference in means is very small relative to the standard deviation, the power is very low.

MINITAB Output

```
Power and Sample Size

2-Sample t Test

Testing mean 1 = mean 2 (versus not =)
Calculating power for mean 1 = mean 2 + difference
Alpha = 0.05   Sigma = 9.32

             Sample
Difference    Size    Power
        2      10    0.0740
```

From the MINITAB output below, the required sample size is 1827. The sample size is huge because the difference in means is very small relative to the standard deviation.

MINITAB Output

```
Power and Sample Size

2-Sample t Test

Testing mean 1 = mean 2 (versus not =)
Calculating power for mean 1 = mean 2 + difference
Alpha = 0.05   Sigma = 9.32

             Sample  Target   Actual
Difference    Size    Power    Power
        1     1827   0.9000   0.9001
```

CHAPTER 3

Experiments with a Single Factor: The Analysis of Variance

LEARNING OBJECTIVES

After completing this chapter, you will be able to:

1. Use the analysis of variance to analyze data from a single-factor completely randomized experiment.

2. Use residual plots to investigate the adequacy of the model and check the validity of the underlying assumptions.

3. Provide a practical interpretation of the results from the experiment.

4. Know how to determine an appropriate sample size for a single-factor fixed effects experiment.

KEY CONCEPTS AND IDEAS

1. Treatments

2. Statistical model

3. Analysis of variance

4. Fixed effects model

5. Residual analysis

6. Variance stabilizing transformation on the response variable

7. Quantitative versus qualitative factors

8. Comparison of individual means following an analysis of variance

9. Operating characteristic curves

Exercises

3.7. The tensile strength of Portland cement is being studied. Four different mixing techniques can be used economically. A completely randomized experiment was conducted and the following data were collected.

Mixing Technique	Tensile Strength (lb/in^2)			
1	3129	3000	2865	2890
2	3200	3300	2975	3150
3	2800	2900	2985	3050
4	2600	2700	2600	2765

(a) Test the hypothesis that mixing techniques affect the strength of the cement. Use $\alpha = 0.05$.

Design-Expert Output

Response: Tensile Strength   in lb/in^2
ANOVA for Selected Factorial Model
Analysis of variance table [Partial sum of squares]

Source	Sum of Squares	DF	Mean Square	F Value	Prob > F	
Model	4.897E+005	3	1.632E+005	12.73	0.0005	significant
A	4.897E+005	3	1.632E+005	12.73	0.0005	
Residual	1.539E+005	12	12825.69			
Lack of Fit	0.000	0				
Pure Error	1.539E+005	12	12825.69			
Cor Total	6.436E+005	15				

The Model F-value of 12.73 implies the model is significant. There is only a 0.05% chance that a "Model F-Value" this large could occur due to noise.

Treatment Means (Adjusted, If Necessary)

	Estimated Mean	Standard Error
1-1	2971.00	56.63
2-2	3156.25	56.63
3-3	2933.75	56.63
4-4	2666.25	56.63

Treatment	Mean Difference	DF	Standard Error	t for H0 Coeff=0	Prob > |t|
1 vs 2	-185.25	1	80.08	-2.31	0.0392
1 vs 3	37.25	1	80.08	0.47	0.6501
1 vs 4	304.75	1	80.08	3.81	0.0025
2 vs 3	222.50	1	80.08	2.78	0.0167
2 vs 4	490.00	1	80.08	6.12	< 0.0001
3 vs 4	267.50	1	80.08	3.34	0.0059

The *F*-value is 12.73 with a corresponding *P*-value of .0005. Mixing technique has an effect.

(b) Construct a graphical display as described in Section 3.5.3 to compare the mean tensile strengths for the four mixing techniques. What are your conclusions?

$$S_{\bar{y}_{i.}} = \sqrt{\frac{MS_E}{n}} = \sqrt{\frac{12825.7}{4}} = 56.625$$

Scaled t Distribution

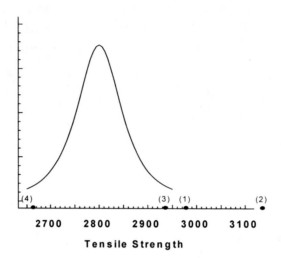

Tensile Strength

Based on examination of the plot, we would conclude that μ_1 and μ_3 are the same; that μ_4 differs from μ_1 and μ_3, that μ_2 differs from μ_1 and μ_3, and that μ_2 and μ_4 are different.

(c) Use the Fisher LSD method with $\alpha=0.05$ to make comparisons between pairs of means.

$$LSD = t_{\alpha/2, N-a}\sqrt{\frac{2MS_E}{n}}$$

$$LSD = t_{0.025, 16-4}\sqrt{\frac{2(12825.7)}{4}}$$

$$LSD = 2.179\sqrt{6412.85} = 174.495$$

> Treatment 2 vs. Treatment 4 = 3156.250 - 2666.250 = 490.000 > 174.495
> Treatment 2 vs. Treatment 3 = 3156.250 - 2933.750 = 222.500 > 174.495
> Treatment 2 vs. Treatment 1 = 3156.250 - 2971.000 = 185.250 > 174.495
> Treatment 1 vs. Treatment 4 = 2971.000 - 2666.250 = 304.750 > 174.495
> Treatment 1 vs. Treatment 3 = 2971.000 - 2933.750 = 37.250 < 174.495
> Treatment 3 vs. Treatment 4 = 2933.750 - 2666.250 = 267.500 > 174.495

The Fisher LSD method is also presented in the Design-Expert computer output above. The results agree with the graphical method for this experiment.

(d) Construct a normal probability plot of the residuals. What conclusion would you draw about the validity of the normality assumption?

There is nothing unusual about the normal probability plot of residuals.

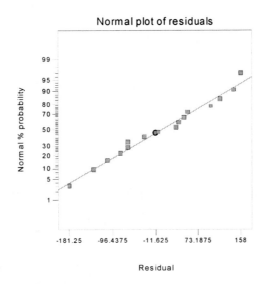

(e) Plot the residuals versus the predicted tensile strength. Comment on the plot.

There is nothing unusual about this plot.

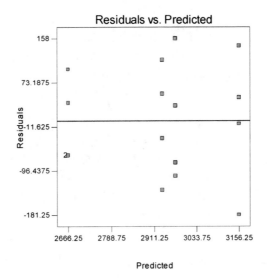

(f) Prepare a scatter plot of the results to aid the interpretation of the results of this experiment.

Design-Expert automatically generates the scatter plot. The plot below also shows the sample average for each treatment and the 95 percent confidence interval on the treatment mean.

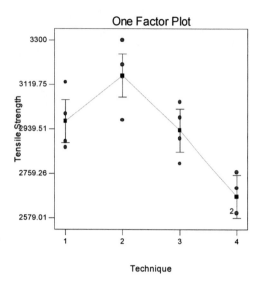

3.9. Reconsider the experiment in Problem 3.7. Find a 95 percent confidence interval on the mean tensile strength of the Portland cement produced by each of the four mixing techniques. Also find a 95 percent confidence interval on the difference in means for techniques 1 and 3. Does this aid in interpreting the results of the experiment?

$$\bar{y}_{i.} - t_{\alpha/2, N-a}\sqrt{\frac{MS_E}{n}} \leq \mu_i \leq \bar{y}_{i.} + t_{\alpha/2, N-a}\sqrt{\frac{MS_E}{n}}$$

Treatment 1: $2971 \pm 2.179\sqrt{\dfrac{12825.69}{4}}$

2971 ± 123.387

$2847.613 \leq \mu_1 \leq 3094.387$

Treatment 2: 3156.25 ± 123.387

$3032.863 \leq \mu_2 \leq 3279.637$

Treatment 3: 2933.75 ± 123.387

$2810.363 \leq \mu_3 \leq 3057.137$

Treatment 4: 2666.25 ± 123.387

$2542.863 \leq \mu_4 \leq 2789.637$

Treatment 1 - Treatment 3: $\bar{y}_{i.} - \bar{y}_{j.} - t_{\alpha/2, N-a}\sqrt{\dfrac{2MS_E}{n}} \leq \mu_i - \mu_j \leq \bar{y}_{i.} - \bar{y}_{j.} + t_{\alpha/2, N-a}\sqrt{\dfrac{2MS_E}{n}}$

$$2971.00 - 2933.75 \pm 2.179\sqrt{\frac{2(12825.7)}{4}}$$

$$-137.245 \leq \mu_1 - \mu_3 \leq 211.745$$

Because the confidence interval for the difference between means 1 and 3 spans zero, we agree with the statement in Problem 3.5 (b); there is not a statistical difference between these two means.

3.10. A product developer is investigating the tensile strength of a new synthetic fiber that will be used to make cloth for men's shirts. Strength is usually affected by the percentage of cotton used in the blend of materials for the fiber. The engineer conducts a completely randomized experiment with five levels of cotton content and replicated the experiment five times. The data are shown in the following table.

Cotton Weight Percentage	Observations				
15	7	7	15	11	9
20	12	17	12	18	18
25	14	19	19	18	18
30	19	25	22	19	23
35	7	10	11	15	11

(a) Is there evidence to support the claim that cotton content affects the mean tensile strength? Use $\alpha = 0.05$.

MINITAB Output

One-way ANOVA: Tensile Strength versus Cotton Percentage

```
Analysis of Variance for Tensile
Source     DF      SS       MS      F        P
Cotton P    4    475.76   118.94   14.76   0.000
Error      20    161.20     8.06
Total      24    636.96
```

Yes, the F-value is 14.76 with a corresponding P-value of 0.000. The percentage of cotton in the fiber appears to have an affect on the tensile strength.

(b) Use the Fisher LSD method to make comparisons between the pairs of means. What conclusions can you draw?

MINITAB Output

```
Fisher's pairwise comparisons

    Family error rate = 0.264
Individual error rate = 0.0500

Critical value = 2.086

Intervals for (column level mean) - (row level mean)

                  15           20           25           30

    20         -9.346
               -1.854

    25        -11.546       -5.946
               -4.054        1.546

    30        -15.546       -9.946       -7.746
               -8.054       -2.454       -0.254

    35         -4.746        0.854        3.054        7.054
                2.746        8.346       10.546       14.546
```

In the MINITAB output the pairs of treatments that do not contain zero in the pair of numbers indicates that there is a difference in the pairs of the treatments. 15% cotton is different than 20%, 25% and 30%. 20% cotton is different than 30% and 35% cotton. 25% cotton is different than 30% and 35% cotton. 30% cotton is different than 35%.

(c) Analyze the residuals from this experiment and comment on model adequacy.

The residual plots below show nothing unusual.

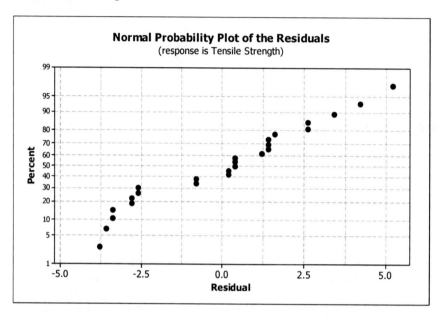

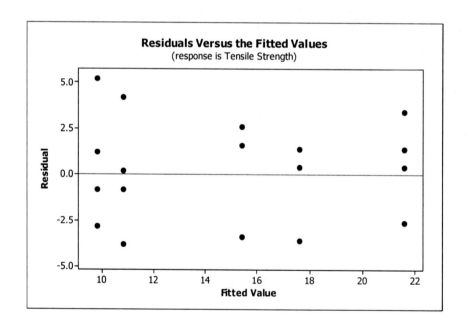

3.12. A pharmaceutical manufacturer wants to investigate the bioactivity of a new drug. A completely randomized single-factor experiment was conducted with three dosage levels, and the following results were obtained.

Dosage	Observations			
20g	24	28	37	30
30g	37	44	31	35
40g	42	47	52	38

(a) Is there evidence to indicate that dosage level affects bioactivity? Use $\alpha = 0.05$.

MINITAB Output

```
One-way ANOVA: Activity versus Dosage

Analysis of Variance for Activity
Source    DF        SS        MS        F        P
Dosage     2     450.7     225.3     7.04    0.014
Error      9     288.3      32.0
Total     11     738.9
```

There appears to be a different in the dosages.

(b) If it is appropriate to do so, make comparisons between the pairs of means. What conclusions can you draw?

Because there appears to be a difference in the dosages, the comparison of means is appropriate.

MINITAB Output

```
Tukey's pairwise comparisons

    Family error rate = 0.0500
Individual error rate = 0.0209

Critical value = 3.95

Intervals for (column level mean) - (row level mean)

              20g           30g

    30g      -18.177
               4.177

    40g      -26.177       -19.177
              -3.823         3.177
```

The Tukey comparison shows a difference in the means between the 20g and the 40g dosages.

(c) Analyze the residuals from this experiment and comment on the model adequacy.

There is nothing too unusual about the residual plots shown below.

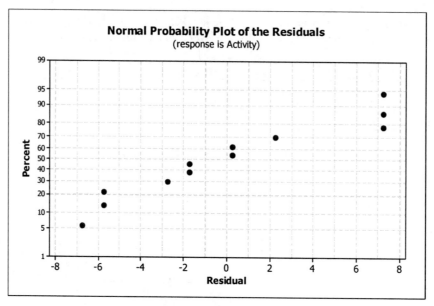

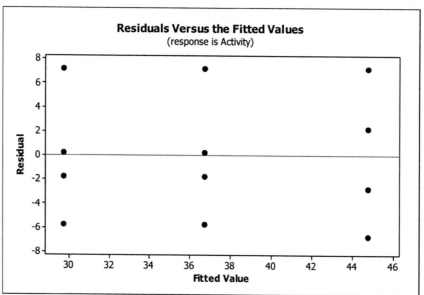

3.14. I belong to a golf club in my neighborhood. I divide the year into three golf seasons: summer (June-September), winter (November-March) and shoulder (October, April and May). I believe that I play my best golf during the summer (because I have more time and the course isn't crowded) and shoulder (because the course isn't crowded) seasons, and my worst golf is during the winter (because when all of the part-year residents show up, and the course is crowded, play is slow, and I get frustrated). Data from the last year are shown in the following table.

Season	Observations									
Summer	83	85	85	87	90	88	88	84	91	90
Shoulder	91	87	84	87	85	86	83			
Winter	94	91	87	85	87	91	92	86		

(a) Do the data indicate that my opinion is correct? Use $\alpha = 0.05$.

MINITAB Output

One-way ANOVA: Score versus Season

```
Analysis of Variance for Score
Source    DF        SS        MS        F        P
Season     2     35.61     17.80     2.12    0.144
Error     22    184.63      8.39
Total     24    220.24
```

The data do not support the author's opinion.

(b) Analyze the residuals from this experiment and comment on model adequacy.

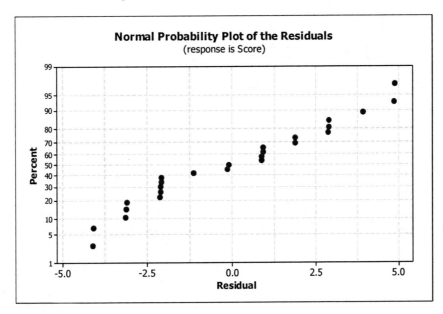

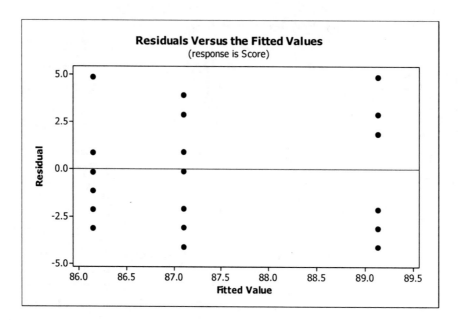

There is nothing unusual about the residuals.

3.18. A manufacturer of television sets is interested in the effect of tube conductivity of four different types of coating for color picture tubes. A completely randomized experiment is conducted and he following conductivity data are obtained:

Coating Type	Conductivity			
1	143	141	150	146
2	152	149	137	143
3	134	136	132	127
4	129	127	132	129

(a) Is there a difference in conductivity due to coating type? Use $\alpha = 0.05$.

Yes, there is a difference in means. Refer to the Design-Expert output below.

Design-Expert Output

ANOVA for Selected Factorial Model
Analysis of variance table [Partial sum of squares]

Source	Sum of Squares	DF	Mean Square	F Value	Prob > F	
Model	844.69	3	281.56	14.30	0.0003	significant
A	844.69	3	281.56	14.30	0.0003	
Residual	236.25	12	19.69			
Lack of Fit	0.000	0				
Pure Error	236.25	12	19.69			
Cor Total	1080.94	15				

The Model F-value of 14.30 implies the model is significant. There is only a 0.03% chance that a "Model F-Value" this large could occur due to noise.

Treatment Means (Adjusted, If Necessary)

	Estimated Mean	Standard Error
1-1	145.00	2.22
2-2	145.25	2.22
3-3	132.25	2.22
4-4	129.25	2.22

Treatment	Mean Difference	DF	Standard Error	t for H0 Coeff=0	Prob > |t|
1 vs 2	-0.25	1	3.14	-0.080	0.9378
1 vs 3	12.75	1	3.14	4.06	0.0016
1 vs 4	15.75	1	3.14	5.02	0.0003
2 vs 3	13.00	1	3.14	4.14	0.0014
2 vs 4	16.00	1	3.14	5.10	0.0003
3 vs 4	3.00	1	3.14	0.96	0.3578

(b) Estimate the overall mean and the treatment effects.

$$\hat{\mu} = 2207 / 16 = 137.9375$$
$$\hat{\tau}_1 = \bar{y}_{1.} - \bar{y}_{..} = 145.00 - 137.9375 = 7.0625$$
$$\hat{\tau}_2 = \bar{y}_{2.} - \bar{y}_{..} = 145.25 - 137.9375 = 7.3125$$
$$\hat{\tau}_3 = \bar{y}_{3.} - \bar{y}_{..} = 132.25 - 137.9375 = -5.6875$$
$$\hat{\tau}_4 = \bar{y}_{4.} - \bar{y}_{..} = 129.25 - 137.9375 = -8.6875$$

(c) Compute a 95 percent interval estimate of the mean of coating type 4. Compute a 99 percent interval estimate of the mean difference between coating types 1 and 4.

$$\text{Treatment 4: } 129.25 \pm 2.179\sqrt{\frac{19.69}{4}}$$
$$124.4155 \le \mu_4 \le 134.0845$$

$$\text{Treatment 1 - Treatment 4: } (145 - 129.25) \pm 3.055\sqrt{\frac{(2)19.69}{4}}$$
$$6.164 \le \mu_1 - \mu_4 \le 25.336$$

(d) Test all pairs of means using the Fisher LSD method with $\alpha = 0.05$.

Refer to the Design-Expert output above. The Fisher LSD procedure is automatically included in the output.

The means of coating type 2 and coating type 1 are not different. The means of coating type 3 and coating type 4 are not different. However, coating types 1 and 2 produce higher mean conductivity than coating types 3 and 4.

(e) Use the graphical method discussed in Section 3.5.3 to compare the means. Which coating type produces the highest conductivity?

$$S_{\bar{y}_i} = \sqrt{\frac{MS_E}{n}} = \sqrt{\frac{19.96}{4}} = 2.219 \quad \text{Coating types 1 and 2 produce the highest conductivity.}$$

Scaled t Distribution

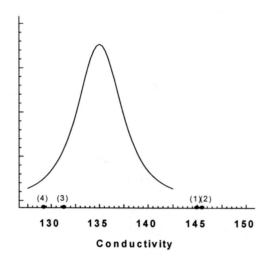

(f) Assuming that coating type 4 is currently in use, what are your recommendations to the manufacturer? We wish to minimize conductivity.

Since coatings 3 and 4 do not differ, and as they both produce the lowest mean values of conductivity, use either coating 3 or 4. As type 4 is currently being used, there is probably no need to change.

3.19. Reconsider the experiment in Problem 3.18. Analyze the residuals and draw conclusions about model adequacy.

There is nothing unusual in the normal probability plot. A funnel shape is seen in the plot of residuals versus predicted conductivity, indicating a possible non-constant variance.

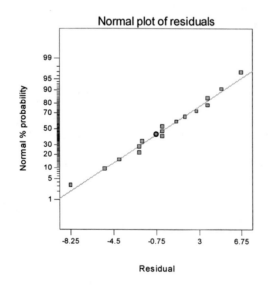

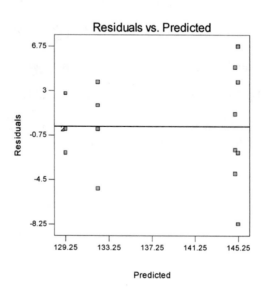

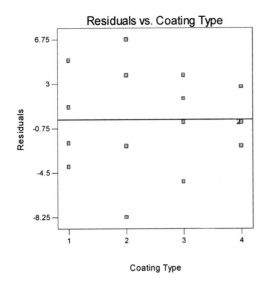

3.21. An article in *Environment International* (Vol. 18, No. 4, 1992) describes an experiment in which the amount of radon released in showers was investigated. Radon-enriched water was used in the experiment and six different orifice diameters were tested in shower heads. The data from the experiment are shown in the following table.

Orifice Dia.	Radon Released (%)			
0.37	80	83	83	85
0.51	75	75	79	79
0.71	74	73	76	77
1.02	67	72	74	74
1.40	62	62	67	69
1.99	60	61	64	66

(a) Does the size of the orifice affect the mean percentage of radon released? Use $\alpha = 0.05$.

Yes. There is at least one treatment mean that is different.

Design-Expert Output

Response: Radon Released in %						
ANOVA for Selected Factorial Model						
Analysis of variance table [Partial sum of squares]						
Source	Sum of Squares	DF	Mean Square	F Value	Prob > F	
Model	1133.38	5	226.68	30.85	< 0.0001	significant
A	1133.38	5	226.68	30.85	< 0.0001	
Residual	132.25	18	7.35			
Lack of Fit	0.000	0				
Pure Error	132.25	18	7.35			
Cor Total	1265.63	23				

The Model F-value of 30.85 implies the model is significant. There is only a 0.01% chance that a "Model F-Value" this large could occur due to noise.

Treatment Means (Adjusted, If Necessary)

	Estimated Standard Mean	Error
1-0.37	82.75	1.36
2-0.51	77.00	1.36
3-0.71	75.00	1.36
4-1.02	71.75	1.36
5-1.40	65.00	1.36
6-1.99	62.75	1.36

Treatment	Mean Difference	DF	Standard Error	t for H0 Coeff=0	Prob > |t|
1 vs 2	5.75	1	1.92	3.00	0.0077
1 vs 3	7.75	1	1.92	4.04	0.0008
1 vs 4	11.00	1	1.92	5.74	< 0.0001
1 vs 5	17.75	1	1.92	9.26	< 0.0001
1 vs 6	20.00	1	1.92	10.43	< 0.0001
2 vs 3	2.00	1	1.92	1.04	0.3105
2 vs 4	5.25	1	1.92	2.74	0.0135
2 vs 5	12.00	1	1.92	6.26	< 0.0001
2 vs 6	14.25	1	1.92	7.43	< 0.0001
3 vs 4	3.25	1	1.92	1.70	0.1072
3 vs 5	10.00	1	1.92	5.22	< 0.0001
3 vs 6	12.25	1	1.92	6.39	< 0.0001
4 vs 5	6.75	1	1.92	3.52	0.0024
4 vs 6	9.00	1	1.92	4.70	0.0002
5 vs 6	2.25	1	1.92	1.17	0.2557

(b) Find the P-value for the F statistic in part (a).

$P = 3.161 \times 10^{-8}$

(c) Analyze the residuals from this experiment.

There is nothing unusual about the residuals.

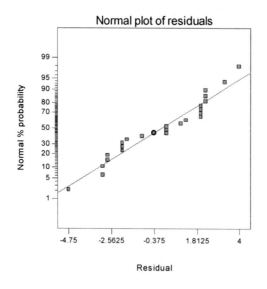

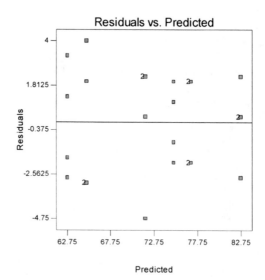

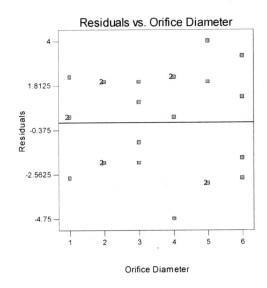

Residuals vs. Orifice Diameter

(d) Find a 95 percent confidence interval on the mean percent radon released when the orifice diameter is 1.40.

Treatment 5 (Orifice =1.40): $65 \pm 2.101 \sqrt{\dfrac{7.35}{4}}$

$62.152 \leq \mu \leq 67.848$

(e) Construct a graphical display to compare the treatment means as described in Section 3-5.3. What conclusions can you draw?

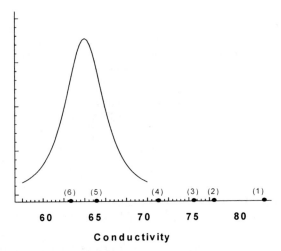

Scaled t Distribution

Treatments 5 and 6 as a group differ from the other means; 2, 3, and 4 as a group differ from the other means; 1 differs from the others.

3.23. The effective life of insulating fluids at an accelerated load of 35 kV is being studied. Test data have been obtained for four types of fluids. The results from a completely randomized experiment were as follows:

Fluid Type	Life (in h) at 35 kV Load					
1	17.6	18.9	16.3	17.4	20.1	21.6
2	16.9	15.3	18.6	17.1	19.5	20.3
3	21.4	23.6	19.4	18.5	20.5	22.3
4	19.3	21.1	16.9	17.5	18.3	19.8

(a) Is there any indication that the fluids differ? Use $\alpha = 0.05$.

At $\alpha = 0.05$ there is no difference, but since the P-value is just slightly above 0.05, there is probably a difference in means.

Design-Expert Output

Response: Life in in h
ANOVA for Selected Factorial Model
Analysis of variance table [Partial sum of squares]

Source	Sum of Squares	DF	Mean Square	F Value	Prob > F	
Model	30.17	3	10.06	3.05	0.0525	not significant
A	30.16	3	10.05	3.05	0.0525	
Residual	65.99	20	3.30			
Lack of Fit	0.000	0				
Pure Error	65.99	20	3.30			
Cor Total	96.16	23				

The Model F-value of 3.05 implies there is a 5.25% chance that a "Model F-Value" this large could occur due to noise.

Treatment Means (Adjusted, If Necessary)

	Estimated Mean	Standard Error
1-1	18.65	0.74
2-2	17.95	0.74
3-3	20.95	0.74
4-4	18.82	0.74

Treatment	Mean Difference	DF	Standard Error	t for H0 Coeff=0	Prob > |t|
1 vs 2	0.70	1	1.05	0.67	0.5121
1 vs 3	-2.30	1	1.05	-2.19	0.0403
1 vs 4	-0.17	1	1.05	-0.16	0.8753
2 vs 3	-3.00	1	1.05	-2.86	0.0097
2 vs 4	-0.87	1	1.05	-0.83	0.4183
3 vs 4	2.13	1	1.05	2.03	0.0554

(b) Which fluid would you select, given that the objective is long life?

Select Treatment 3. The Fisher LSD procedure in the computer output indicates that fluid 3 is different from the others, and its average life also exceeds the average lives of the other three fluids.

(c) Analyze the residuals from this experiment. Are the basic analysis of variance assumptions satisfied?

There is nothing unusual in the residual plots shown.

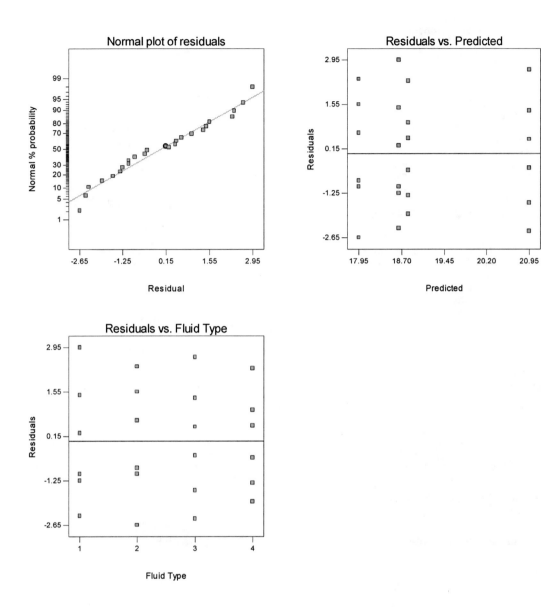

3.26. Three brands of batteries are under study. It is suspected that the lives (in weeks) of the three brands are different. Five randomly selected batteries of each brand are tested with the following results:

	Weeks of Life	
Brand 1	Brand 2	Brand 3
100	76	108
96	80	100
92	75	96
96	84	98
92	82	100

(a) Are the lives of these brands of batteries different?

Yes, at least one of the brands is different.

Design-Expert Output

Response: Life in Weeks
　　　ANOVA for Selected Factorial Model
Analysis of variance table [Partial sum of squares]

Source	Sum of Squares	DF	Mean Square	F Value	Prob > F	
Model	1196.13	2	598.07	38.34	< 0.0001	significant
A	*1196.13*	*2*	*598.07*	*38.34*	*< 0.0001*	
Residual	187.20	12	15.60			
Lack of Fit	*0.000*	*0*				
Pure Error	*187.20*	*12*	*15.60*			
Cor Total	1383.33	14				

The Model F-value of 38.34 implies the model is significant. There is only
a 0.01% chance that a "Model F-Value" this large could occur due to noise.

Treatment Means (Adjusted, If Necessary)

	Estimated Mean	Standard Error
1-1	95.20	1.77
2-2	79.40	1.77
3-3	100.40	1.77

Treatment	Mean Difference	DF	Standard Error	t for H0 Coeff=0	Prob > |t|
1 vs 2	15.80	1	2.50	6.33	< 0.0001
1 vs 3	-5.20	1	2.50	-2.08	0.0594
2 vs 3	-21.00	1	2.50	-8.41	< 0.0001

(b) Analyze the residuals from this experiment.

There is nothing unusual about the residuals.

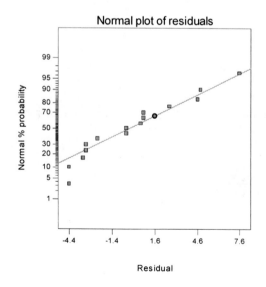

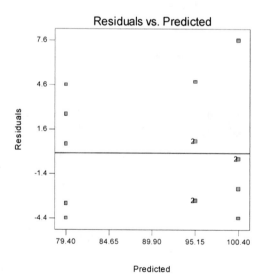

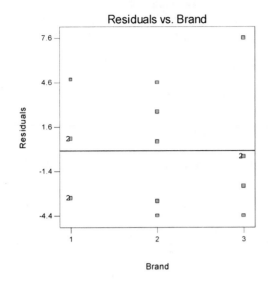

(c) Construct a 95 percent confidence interval estimate on the mean life of battery brand 2. Construct a 99 percent confidence interval estimate on the mean difference between the lives of battery brands 2 and 3.

$$\bar{y}_{i.} \pm t_{\alpha/2, N-a}\sqrt{\frac{MS_E}{n}}$$

Brand 2: $79.4 \pm 2.179\sqrt{\dfrac{15.60}{5}}$

$$79.40 \pm 3.849$$
$$75.551 \le \mu_2 \le 83.249$$

Brand 2 - Brand 3: $\bar{y}_{i.} - \bar{y}_{j.} \pm t_{\alpha/2, N-a}\sqrt{\dfrac{2MS_E}{n}}$

$$79.4 - 100.4 \pm 3.055\sqrt{\frac{2(15.60)}{5}}$$

$$-28.631 \le \mu_2 - \mu_3 \le -13.369$$

(d) Which brand would you select for use? If the manufacturer will replace without charge any battery that fails in less than 85 weeks, what percentage would the company expect to replace?

Choose brand 3 for longest life. Mean life of this brand in 100.4 weeks, and the variance of life is estimated by 15.60 (*MSE*). Assuming normality, then the probability of failure before 85 weeks is:

$$\Phi\left(\frac{85 - 100.4}{\sqrt{15.60}}\right) = \Phi(-3.90) = 0.00005$$

That is, about 5 out of 100,000 batteries will fail before 85 weeks.

3.28. An experiment was performed to investigate the effectiveness of five insulating materials. Four samples of each material were tested at an elevated voltage level to accelerate the time to failure. The failure times (in minutes) are shown below:

Material	Failure Time (minutes)			
1	110	157	194	178
2	1	2	4	18
3	880	1256	5276	4355
4	495	7040	5307	10050
5	7	5	29	2

(a) Do all five materials have the same effect on mean failure time?

No, at least one material is different.

Design-Expert Output

Response: Failure Time   in Minutes
ANOVA for Selected Factorial Model
Analysis of variance table [Partial sum of squares]

Source	Sum of Squares	DF	Mean Square	F Value	Prob > F	
Model	1.032E+008	4	2.580E+007	6.19	0.0038	significant
A	1.032E+008	4	2.580E+007	6.19	0.0038	
Residual	6.251E+007	15	4.167E+006			
Lack of Fit	0.000	0				
Pure Error	6.251E+007	15	4.167E+006			
Cor Total	1.657E+008	19				

The Model F-value of 6.19 implies the model is significant. There is only a 0.38% chance that a "Model F-Value" this large could occur due to noise.

Treatment Means (Adjusted, If Necessary)

	Estimated Mean	Standard Error
1-1	159.75	1020.67
2-2	6.25	1020.67
3-3	2941.75	1020.67
4-4	5723.00	1020.67
5-5	10.75	1020.67

Treatment	Mean Difference	DF	Standard Error	t for H0 Coeff=0	Prob > |t|
1 vs 2	153.50	1	1443.44	0.11	0.9167
1 vs 3	-2782.00	1	1443.44	-1.93	0.0731
1 vs 4	-5563.25	1	1443.44	-3.85	0.0016
1 vs 5	149.00	1	1443.44	0.10	0.9192
2 vs 3	-2935.50	1	1443.44	-2.03	0.0601
2 vs 4	-5716.75	1	1443.44	-3.96	0.0013
2 vs 5	-4.50	1	1443.44	-3.118E-003	0.9976
3 vs 4	-2781.25	1	1443.44	-1.93	0.0732
3 vs 5	2931.00	1	1443.44	2.03	0.0604
4 vs 5	5712.25	1	1443.44	3.96	0.0013

(b) Plot the residuals versus the predicted response. Construct a normal probability plot of the residuals. What information is conveyed by these plots?

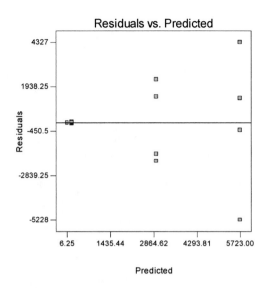

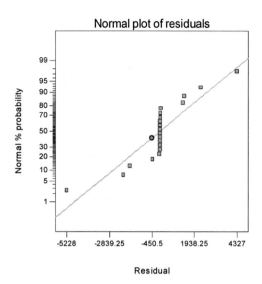

The plot of residuals versus predicted has a strong outward-opening funnel shape, which indicates the variance of the original observations is not constant. The normal probability plot also indicates that the normality assumption is not valid. A data transformation is recommended.

(c) Based on your answer to part (b), conduct another analysis of the failure time data and draw appropriate conclusions.

A natural log transformation was applied to the failure time data. The analysis in the log scale identifies that there exists at least one difference in treatment means.

Design-Expert Output

Response: Failure Time in Minutes   Transform: Natural log Constant: 0.000
ANOVA for Selected Factorial Model
Analysis of variance table [Partial sum of squares]

Source	Sum of Squares	DF	Mean Square	F Value	Prob > F	
Model	165.06	4	41.26	37.66	< 0.0001	significant
A	165.06	4	41.26	37.66	< 0.0001	
Residual	16.44	15	1.10			
Lack of Fit	0.000	0				
Pure Error	16.44	15	1.10			
Cor Total	181.49	19				

The Model F-value of 37.66 implies the model is significant. There is only a 0.01% chance that a "Model F-Value" this large could occur due to noise.

Treatment Means (Adjusted, If Necessary)

	Estimated Mean	Standard Error
1-1	5.05	0.52
2-2	1.24	0.52
3-3	7.72	0.52
4-4	8.21	0.52
5-5	1.90	0.52

Treatment	Mean Difference	DF	Standard Error	t for H0 Coeff=0	Prob > |t|
1 vs 2	3.81	1	0.74	5.15	0.0001
1 vs 3	-2.66	1	0.74	-3.60	0.0026
1 vs 4	-3.16	1	0.74	-4.27	0.0007
1 vs 5	3.15	1	0.74	4.25	0.0007
2 vs 3	-6.47	1	0.74	-8.75	< 0.0001
2 vs 4	-6.97	1	0.74	-9.42	< 0.0001
2 vs 5	-0.66	1	0.74	-0.89	0.3856
3 vs 4	-0.50	1	0.74	-0.67	0.5116
3 vs 5	5.81	1	0.74	7.85	< 0.0001
4 vs 5	6.31	1	0.74	8.52	< 0.0001

There is nothing unusual about the residual plots when the natural log transformation is applied.

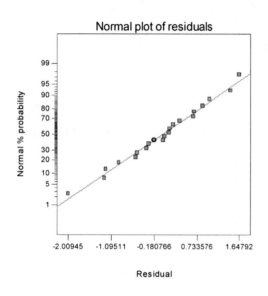

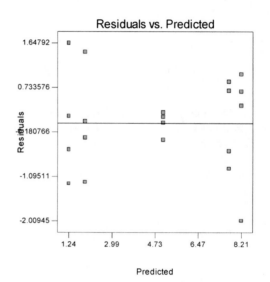

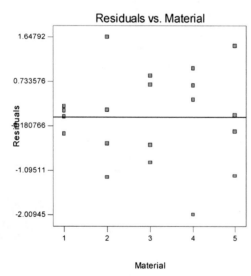

3.42. Refer to Problem 3.26.

(a) If we wish to detect a maximum difference in mean battery life of 10 hours with a probability of at least 0.90, what sample size should be used? Discuss how you would obtain a preliminary estimate of σ^2 for answering this question.

Use the MS_E from Problem 3.26.

$$\Phi^2 = \frac{nD^2}{2a\sigma^2} \qquad \Phi^2 = \frac{n(10)^2}{2(3)(15.60)} = 1.0684n$$

Letting $\alpha = 0.05$, P(accept) = 0.1, $\upsilon_1 = a - 1 = 2$

Trial and Error yields:

n	υ_2	Φ	P(accept)
4	9	2.067	0.25
5	12	2.311	0.12
6	15	2.532	0.05

Choose n ≥ 6, therefore N ≥ 18

See the discussion from the previous problem about the estimate of variance.

(b) If the difference between brands is great enough so that the standard deviation of an observation is increased by 25 percent, what sample size should be used if we wish to detect this with a probability of at least 0.90?

$$\upsilon_1 = a - 1 = 2 \qquad \upsilon_2 = N - a = 15 - 3 = 12 \qquad \alpha = 0.05 \qquad P(accept) \le 0.1$$

$$\lambda = \sqrt{1 + n\left[(1 + 0.01P)^2 - 1\right]} = \sqrt{1 + n\left[(1 + 0.01(25))^2 - 1\right]} = \sqrt{1 + 0.5625n}$$

Trial and Error yields:

n	υ_2	Φ	P(accept)
8	21	2.12	0.16
9	24	2.25	0.13
10	27	2.37	0.09

Choose n ≥ 10, therefore N ≥ 30

3.44. Suppose that four normal populations have means of $\mu_1{=}50$, $\mu_2{=}60$, $\mu_3{=}50$, and $\mu_4{=}60$. How many observations should be taken from each population so that the probability of rejecting the null hypothesis of equal population means is at least 0.90? Assume that $\alpha{=}0.05$ and that a reasonable estimate of the error variance is $\sigma^2{=}25$.

$$\mu_i = \mu + \tau_i, i = 1,2,3,4$$

$$\mu = \frac{\sum_{i=1}^{4} \mu_i}{4} = \frac{220}{4} = 55 \qquad \Phi^2 = \frac{n\sum \tau_i^2}{a\sigma^2} = \frac{100n}{4(25)} = n$$

$$\tau_1 = -5, \tau_2 = 5, \tau_3 = -5, \tau_4 = 5 \qquad \Phi = \sqrt{n}$$

$$\sum_{i=1}^{4} \tau_i^2 = 100$$

$\upsilon_1 = 3, \upsilon_2 = 4(n-1), \alpha = 0.05$, from the O.C. curves we can construct the following:

n	Φ	υ_2	β	1-β
4	2.00	12	0.18	0.82
5	2.24	16	0.08	0.92

Therefore, select n=5

3.47. Refer to the aluminum smelting experiment in Section 3.8.3. Verify the ANOVA for pot noise summarized in Table 3.16. Examine the usual residual plots and comment on the experimental validity.

Design-Expert Output

Response: Cell StDev   Transform: Natural log   Constant: 0.000
ANOVA for Selected Factorial Model
Analysis of variance table [Partial sum of squares]

Source	Sum of Squares	DF	Mean Square	F Value	Prob > F	
Model	6.17	3	2.06	21.96	< 0.0001	significant
A	6.17	3	2.06	21.96	< 0.0001	
Residual	1.87	20	0.094			
Lack of Fit	0.000	0				
Pure Error	1.87	20	0.094			
Cor Total	8.04	23				

The Model F-value of 21.96 implies the model is significant. There is only a 0.01% chance that a "Model F-Value" this large could occur due to noise.

Treatment Means (Adjusted, If Necessary)

	Estimated Mean	Standard Error
1-1	-3.09	0.12
2-2	-3.51	0.12
3-3	-2.20	0.12
4-4	-3.36	0.12

Treatment	Mean Difference	DF	Standard Error	t for H0 Coeff=0	Prob > |t|
1 vs 2	0.42	1	0.18	2.38	0.0272
1 vs 3	-0.89	1	0.18	-5.03	< 0.0001
1 vs 4	0.27	1	0.18	1.52	0.1445
2 vs 3	-1.31	1	0.18	-7.41	< 0.0001
2 vs 4	-0.15	1	0.18	-0.86	0.3975
3 vs 4	1.16	1	0.18	6.55	< 0.0001

The following residual plots identify the residuals to be normally distributed, randomly distributed through the range of prediction, and uniformly distributed across the different algorithms. This validates the assumptions for the experiment.

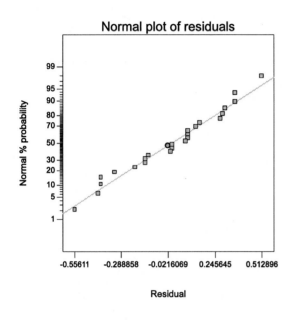

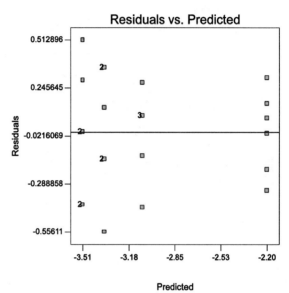

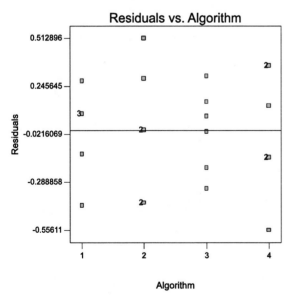

3.52. Use the Kruskal-Wallis test for the experiment in Problem 3.23. Compare the conclusions obtained with those from the usual analysis of variance.

From Design-Expert Output of Problem 3.23

Response: Life in in h
ANOVA for Selected Factorial Model
Analysis of variance table [Partial sum of squares]

Source	Sum of Squares	DF	Mean Square	F Value	Prob > F	
Model	30.17	3	10.06	3.05	0.0525	not significant
A	30.16	3	10.05	3.05	0.0525	
Residual	65.99	20	3.30			
Lack of Fit	0.000	0				
Pure Error	65.99	20	3.30			
Cor Total	96.16	23				

$$H = \frac{12}{N(N+1)}\left[\sum_{i=1}^{a}\frac{R_i^2}{n_i}\right] - 3(N+1) = \frac{12}{24(24+1)}[4060.75] - 3(24+1) = 6.22$$

$$\chi_{0.05,3}^2 = 7.81$$

Accept the null hypothesis; the treatments are not different. This agrees with the analysis of variance.

CHAPTER 4

Randomized Blocks, Latin Squares, and Related Designs

LEARNING OBJECTIVES

After completing this chapter, you will be able to:

1. Design, conduct, and analyze experiments using the randomized complete block design (RCBD).

2. Use residual analysis to investigate the adequacy of the RCBD model and check the validity of the underlying assumptions.

3. Design, conduct, and analyze experiments using the Latin square design.

4. Explain how the blocking principle can be used to eliminate known and controllable sources of nuisance variation in an experiment.

KEY CONCEPTS AND IDEAS

1. Nuisance factor

2. Randomized complete block design (RCBD)

3. Statistical model for the RCBD

4. Analysis of variance for the RCBD

5. Model adequacy checking for the RCBD

6. Latin square design

7. Graeco-Latin square design

8. Balanced incomplete block design

Exercises

4.4.   Three different washing solutions are being compared to study their effectiveness in retarding bacteria growth in five-gallon milk containers. The analysis is done in a laboratory, and only three trials can be run on any day. Because days could represent a potential source of variability, the experimenter decides to use a randomized block design. Observations are taken for four days, and the data are shown here. Analyze the data from this experiment (use $\alpha = 0.05$) and draw conclusions.

	Days			
Solution	1	2	3	4
1	13	22	18	39
2	16	24	17	44
3	5	4	1	22

Design-Expert Output

Response: Growth
   ANOVA for Selected Factorial Model
Analysis of variance table [Partial sum of squares]

Source	Sum of Squares	DF	Mean Square	F Value	Prob > F	
Block	1106.92	3	368.97			
Model	703.50	2	351.75	40.72	0.0003	significant
A	703.50	2	351.75	40.72	0.0003	
Residual	51.83	6	8.64			
Cor Total	1862.25	11				

The Model F-value of 40.72 implies the model is significant. There is only a 0.03% chance that a "Model F-Value" this large could occur due to noise.

Std. Dev.	2.94	R-Squared	0.9314	
Mean	18.75	Adj R-Squared	0.9085	
C.V.	15.68	Pred R-Squared	0.7255	
PRESS	207.33	Adeq Precision	19.687	

Treatment Means (Adjusted, If Necessary)

	Estimated Mean	Standard Error
1-1	23.00	1.47
2-2	25.25	1.47
3-3	8.00	1.47

Treatment	Mean Difference	DF	Standard Error	t for H0 Coeff=0	Prob > |t|
1 vs 2	-2.25	1	2.08	-1.08	0.3206
1 vs 3	15.00	1	2.08	7.22	0.0004
2 vs 3	17.25	1	2.08	8.30	0.0002

There is a difference between the means of the three solutions. The Fisher LSD procedure indicates that solution 3 is significantly different than the other two.

4.7. Consider the hardness testing experiment described in Section 4.1. Suppose that the experiment was conducted as described and the following Rockwell C-scale data (coded by subtracting 40 units) obtained:

	Coupon			
Tip	1	2	3	4
1	9.3	9.4	9.6	10.0
2	9.4	9.3	9.8	9.9
3	9.2	9.4	9.5	9.7
4	9.7	9.6	10.0	10.2

(a) Analyze the data from this experiment.

There is a difference between the means of the four tips.

Design-Expert Output

Response: Hardness
ANOVA for Selected Factorial Model
Analysis of variance table [Terms added sequentially (first to last)]

Source	Sum of Squares	DF	Mean Square	F Value	Prob > F	
Block	0.82	3	0.27			
Model	0.38	3	0.13	14.44	0.0009	significant
A	0.38	3	0.13	14.44	0.0009	
Residual	0.080	9	8.889E-003			
Cor Total	1.29	15				

The Model F-value of 14.44 implies the model is significant. There is only a 0.09% chance that a "Model F-Value" this large could occur due to noise.

Std. Dev.	0.094	R-Squared	0.8280
Mean	9.63	Adj R-Squared	0.7706
C.V.	0.98	Pred R-Squared	0.4563
PRESS	0.25	Adeq Precision	15.635

Treatment Means (Adjusted, If Necessary)

	Estimated Mean	Standard Error
1-1	9.57	0.047
2-2	9.60	0.047
3-3	9.45	0.047
4-4	9.88	0.047

Treatment	Mean Difference	DF	Standard Error	t for H0 Coeff=0	Prob > |t|
1 vs 2	-0.025	1	0.067	-0.38	0.7163
1 vs 3	0.13	1	0.067	1.87	0.0935
1 vs 4	-0.30	1	0.067	-4.50	0.0015
2 vs 3	0.15	1	0.067	2.25	0.0510
2 vs 4	-0.27	1	0.067	-4.12	0.0026
3 vs 4	-0.43	1	0.067	-6.37	0.0001

(b) Use the Fisher LSD method to make comparisons among the four tips to determine specifically which tips differ in mean hardness readings.

Based on the LSD bars in the Design-Expert plot that follows, the mean of tip 4 differs from the means of tips 1, 2, and 3. The LSD method identifies a marginal difference between the means of tips 2 and 3.

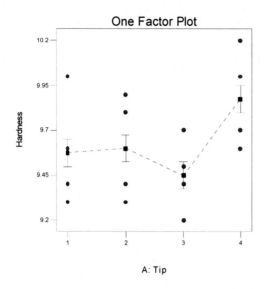

(c) Analyze the residuals from this experiment.

The residual plots below do not identify any violations to the assumptions.

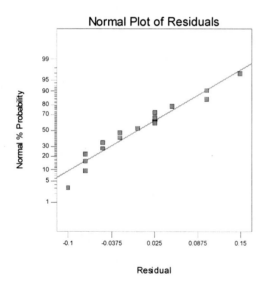

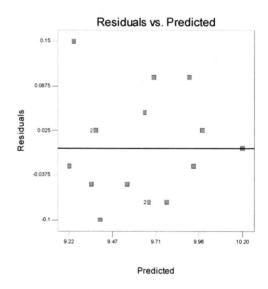

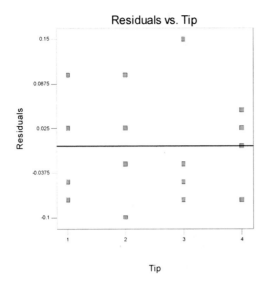

Residuals vs. Tip

4.8. A consumer products company relies on direct mail marketing pieces as a major component of its advertising campaigns. The company has three different designs for a new brochure and want to evaluate their effectiveness, as there are substantial differences in costs between the three designs. The company decides to test the three designs by mailing 5,000 samples of each to potential customers in four different regions of the country. Since there are known regional differences in the customer base, regions are considered as blocks. The number of responses to each mailing is shown below.

Design	Region			
	NE	NW	SE	SW
1	250	350	219	375
2	400	525	390	580
3	275	340	200	310

(a) Analyze the data from this experiment.

The residuals of the analysis below identify concerns with the normality and equality of variance assumptions. As a result, a square root transformation was applied as shown in the second ANOVA table. The residuals of both analyses are presented for comparison in part (c) of this problem. The analysis concludes that there is a difference between the mean number of responses for the three designs.

Design-Expert Output

Response: Number of responses
ANOVA for Selected Factorial Model
Analysis of variance table [Terms added sequentially (first to last)]

Source	Sum of Squares	DF	Mean Square	F Value	Prob > F	
Block	49035.67	3	16345.22			
Model	90755.17	2	45377.58	50.15	0.0002	significant
A	90755.17	2	45377.58	50.15	0.0002	
Residual	5428.83	6	904.81			
Cor Total	1.452E+005	11				

The Model F-value of 50.15 implies the model is significant. There is only a 0.02% chance that a "Model F-Value" this large could occur due to noise.

Std. Dev.	30.08	R-Squared	0.9436	
Mean	351.17	Adj R-Squared	0.9247	
C.V.	8.57	Pred R-Squared	0.7742	
PRESS	21715.33	Adeq Precision	16.197	

Treatment Means (Adjusted, If Necessary)

	Estimated Mean	Standard Error
1-1	298.50	15.04
2-2	473.75	15.04
3-3	281.25	15.04

Treatment	Mean Difference	DF	Standard Error	t for H0 Coeff=0	Prob > |t|
1 vs 2	-175.25	1	21.27	-8.24	0.0002
1 vs 3	17.25	1	21.27	0.81	0.4483
2 vs 3	192.50	1	21.27	9.05	0.0001

Design-Expert Output for Transformed Data

Response: Number of responses   **Transform: Square root**   **Constant: 0**
ANOVA for Selected Factorial Model
Analysis of variance table [Terms added sequentially (first to last)]

Source	Sum of Squares	DF	Mean Square	F Value	Prob > F	
Block	35.89	3	11.96			
Model	60.73	2	30.37	60.47	0.0001	significant
A	60.73	2	30.37	60.47	0.0001	
Residual	3.01	6	0.50			
Cor Total	99.64	11				

The Model F-value of 60.47 implies the model is significant. There is only a 0.01% chance that a "Model F-Value" this large could occur due to noise.

Std. Dev.	0.71	R-Squared	0.9527	
Mean	18.52	Adj R-Squared	0.9370	
C.V.	3.83	Pred R-Squared	0.8109	
PRESS	12.05	Adeq Precision	18.191	

Treatment Means (Adjusted, If Necessary)

	Estimated Mean	Standard Error
1-1	17.17	0.35
2-2	21.69	0.35
3-3	16.69	0.35

Treatment	Mean Difference	DF	Standard Error	t for H0 Coeff=0	Prob > |t|
1 vs 2	-4.52	1	0.50	-9.01	0.0001
1 vs 3	0.48	1	0.50	0.95	0.3769
2 vs 3	4.99	1	0.50	9.96	< 0.0001

(b) Use the Fisher LSD method to make comparisons among the three designs to determine specifically which designs differ in mean response rate.

Based on the LSD bars in the Design-Expert plot that follows, designs 1 and 3 do not differ; however, design 2 is different than designs 1 and 3.

One Factor Plot

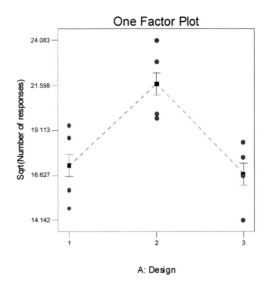

(c) Analyze the residuals from this experiment.

The first set of residual plots presented below represent the untransformed data. Concerns with normality as well as inequality of variance are presented. The second set of residual plots represents transformed data and does not identify significant violations of the assumptions. The residuals vs. design plot indicates a slight inequality of variance; however, it is not a strong violation and is an improvement over the non-transformed data.

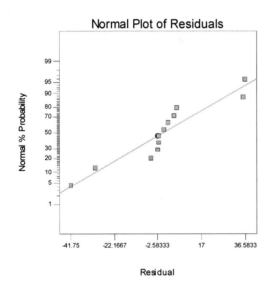

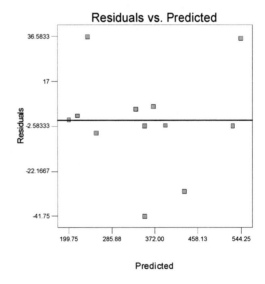

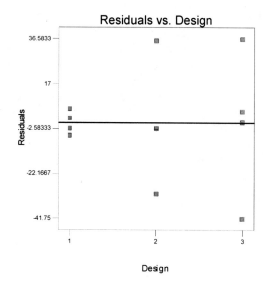

The following are the square root transformed data residual plots.

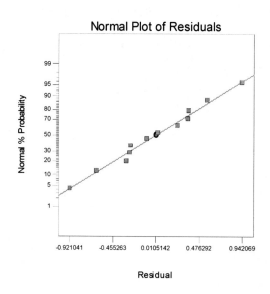

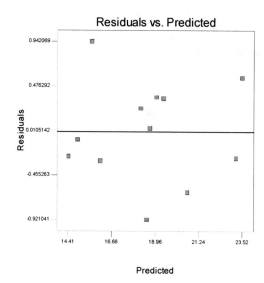

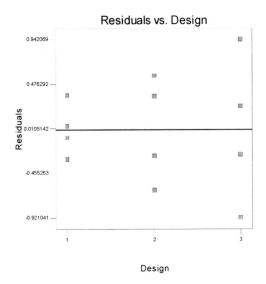

Residuals vs. Design

4.14. An aluminum master alloy manufacturer produces grain refiners in ingot form. The company produces the product in four furnaces. Each furnace is known to have its own unique operating characteristics, so any experiment run in the foundry that involves more than one furnace will consider furnaces as a nuisance variable. The process engineers suspect that stirring rate impacts the grain size of the product. Each furnace can be run at four different stirring rates. A randomized block design is run for a particular refiner and the resulting grain size data is as follows.

	Furnace			
Stirring Rate	1	2	3	4
5	8	4	5	6
10	14	5	6	9
15	14	6	9	2
20	17	9	3	6

(a) Is there any evidence that stirring rate impacts grain size?

Design-Expert Output

Response: Grain Size
ANOVA for Selected Factorial Model
Analysis of variance table [Partial sum of squares]

Source	Sum of Squares	DF	Mean Square	F Value	Prob > F	
Block	165.19	3	55.06			
Model	22.19	3	7.40	0.85	0.4995	not significant
A	22.19	3	7.40	0.85	0.4995	
Residual	78.06	9	8.67			
Cor Total	265.44	15				

The "Model F-value" of 0.85 implies the model is not significant relative to the noise. There is a 49.95 % chance that a "Model F-value" this large could occur due to noise.

Std. Dev.	2.95	R-Squared	0.2213
Mean	7.69	Adj R-Squared	-0.0382
C.V.	38.31	Pred R-Squared	-1.4610
PRESS	246.72	Adeq Precision	5.390

Treatment Means (Adjusted, If Necessary)

	Estimated Mean	Standard Error
1-5	5.75	1.47
2-10	8.50	1.47
3-15	7.75	1.47
4-20	8.75	1.47

Treatment	Mean Difference	DF	Standard Error	t for H$_0$ Coeff=0	Prob > |t|
1 vs 2	-2.75	1	2.08	-1.32	0.2193
1 vs 3	-2.00	1	2.08	-0.96	0.3620
1 vs 4	-3.00	1	2.08	-1.44	0.1836
2 vs 3	0.75	1	2.08	0.36	0.7270
2 vs 4	-0.25	1	2.08	-0.12	0.9071
3 vs 4	-1.00	1	2.08	-0.48	0.6425

The analysis of variance shown above indicates that there is no difference in mean grain size due to the different stirring rates.

(b) Graph the residuals from this experiment on a normal probability plot. Interpret this plot.

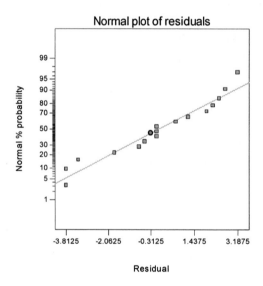

The plot indicates that the normality assumption is valid.

(c) Plot the residuals versus furnace and stirring rate. Does this plot convey any useful information?

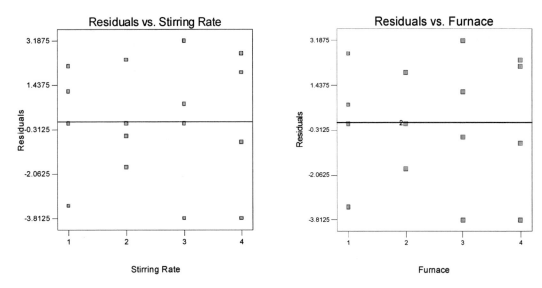

The variance is consistent at different stirring rates. Not only does this validate the assumption of uniform variance, it also identifies that the different stirring rates do not affect variance.

(d) What should the process engineers recommend concerning the choice of stirring rate and furnace for this particular grain refiner if small grain size is desirable?

There really is no effect due to the stirring rate.

4.15. Analyze the data in Problem 4.4 using the general regression significance test.

$$
\begin{aligned}
\mu: &\quad 12\hat{\mu} &&+4\hat{\tau}_1 &&+4\hat{\tau}_2 &&+4\hat{\tau}_3 &&+3\hat{\beta}_1 &&+3\hat{\beta}_2 &&+3\hat{\beta}_3 &&+3\hat{\beta}_4 &&=225 \\
\tau_1: &\quad 4\hat{\mu} &&+4\hat{\tau}_1 && && &&+\hat{\beta}_1 &&+\hat{\beta}_2 &&+\hat{\beta}_3 &&+\hat{\beta}_4 &&=92 \\
\tau_2: &\quad 4\hat{\mu} && &&+4\hat{\tau}_2 && &&+\hat{\beta}_1 &&+\hat{\beta}_2 &&+\hat{\beta}_3 &&+\hat{\beta}_4 &&=101 \\
\tau_3: &\quad 4\hat{\mu} && && &&+4\hat{\tau}_3 &&+\hat{\beta}_1 &&+\hat{\beta}_2 &&+\hat{\beta}_3 &&+\hat{\beta}_4 &&=32 \\
\beta_1: &\quad 3\hat{\mu} &&+\hat{\tau}_1 &&+\hat{\tau}_2 &&+\hat{\tau}_3 &&+3\hat{\beta}_1 && && && &&=34 \\
\beta_1: &\quad 3\hat{\mu} &&+\hat{\tau}_1 &&+\hat{\tau}_2 &&+\hat{\tau}_3 && &&+3\hat{\beta}_2 && && &&=50 \\
\beta_1: &\quad 3\hat{\mu} &&+\hat{\tau}_1 &&+\hat{\tau}_2 &&+\hat{\tau}_3 && && &&+3\hat{\beta}_3 && &&=36 \\
\beta: &\quad 3\hat{\mu} &&+\hat{\tau}_1 &&+\hat{\tau}_2 &&+\hat{\tau}_3 && && && &&+3\hat{\beta}_4 &&=105
\end{aligned}
$$

Applying the constraints $\sum \hat{\tau}_i = \sum \hat{\beta}_j = 0$, we obtain:

$$
\hat{\mu} = \frac{225}{12},\ \hat{\tau}_1 = \frac{51}{12},\ \hat{\tau}_2 = \frac{78}{12},\ \hat{\tau}_3 = \frac{-129}{12},\ \hat{\beta}_1 = \frac{-89}{12},\ \hat{\beta}_2 = \frac{-25}{12},\ \hat{\beta}_3 = \frac{-81}{12},\ \hat{\beta}_4 = \frac{195}{12}
$$

$$R(\mu,\tau,\beta) = \left(\frac{225}{12}\right)(225) + \left(\frac{51}{12}\right)(92) + \left(\frac{78}{12}\right)(101) + \left(\frac{-129}{12}\right)(32) + \left(\frac{-89}{12}\right)(34) + \left(\frac{-25}{12}\right)(50) +$$

$$\left(\frac{-81}{12}\right)(36) + \left(\frac{195}{12}\right)(105) = 6029.17$$

$$\sum\sum y_{ij}^2 = 6081, \ SS_E = \sum\sum y_{ij}^2 - R(\mu,\tau,\beta) = 6081 - 6029.17 = 51.83$$

Model Restricted to $\tau_i = 0$:

$$
\begin{aligned}
\mu: \quad & 12\hat{\mu} \ +3\hat{\beta}_1 \ +3\hat{\beta}_2 \ +3\hat{\beta}_3 \ +3\hat{\beta}_4 \ = 225 \\
\beta_1: \quad & 3\hat{\mu} \ +3\hat{\beta}_1 \ &= 34 \\
\beta_2: \quad & 3\hat{\mu} \ +3\hat{\beta}_2 \ &= 50 \\
\beta_3: \quad & 3\hat{\mu} \ +3\hat{\beta}_3 \ &= 36 \\
\beta_4: \quad & 3\hat{\mu} \ +3\hat{\beta}_4 \ = 105
\end{aligned}
$$

Applying the constraint $\sum\hat{\beta}_j = 0$, we obtain:

$$\hat{\mu} = \frac{225}{12} \ , \ \hat{\beta}_1 = -89/12, \ \hat{\beta}_2 = \frac{-25}{12}, \ \hat{\beta}_3 = \frac{-81}{12}, \ \hat{\beta}_4 = \frac{195}{12}. \ \text{Now:}$$

$$R(\mu,\beta) = \left(\frac{225}{12}\right)(225) + \left(\frac{-89}{12}\right)(34) + \left(\frac{-25}{12}\right)(50) + \left(\frac{-81}{12}\right)(36) + \left(\frac{195}{12}\right)(105) = 5325.67$$

$$R(\tau|\mu,\beta) = R(\mu,\tau,\beta) - R(\mu,\beta) = 6029.17 - 5325.67 = 703.50 = SS_{Treatments}$$

Model Restricted to $\beta_j = 0$:

$$
\begin{aligned}
\mu: \quad & 12\hat{\mu} \ +4\hat{\tau}_1 \ +4\hat{\tau}_2 \ +4\hat{\tau}_3 \ = 225 \\
\tau_1: \quad & 4\hat{\mu} \ +4\hat{\tau}_1 \ &= 92 \\
\tau_2: \quad & 4\hat{\mu} \ +4\hat{\tau}_2 \ &= 101 \\
\tau_3: \quad & 4\hat{\mu} \ +4\hat{\tau}_3 \ = 32
\end{aligned}
$$

Applying the constraint $\sum\hat{\tau}_i = 0$, we obtain:

$$\hat{\mu} = \frac{225}{12}, \ \hat{\tau}_1 = \frac{51}{12}, \ \hat{\tau}_2 = \frac{78}{12}, \ \hat{\tau}_3 = \frac{-129}{12}$$

$$R(\mu,\tau) = \left(\frac{225}{12}\right)(225) + \left(\frac{51}{12}\right)(92) + \left(\frac{78}{12}\right)(101) + \left(\frac{-129}{12}\right)(32) = 4922.25$$

$$R(\beta|\mu,\tau) = R(\mu,\tau,\beta) - R(\mu,\tau) = 6029.17 - 4922.25 = 1106.92 = SS_{Blocks}$$

4.16. Assuming that chemical types and bolts are fixed, estimate the model parameters τ_i and β_j in Problem 4.3.

Using Equations 4-18, applying the constraints, we obtain:

$$\hat{\mu} = \frac{35}{20}, \ \hat{\tau}_1 = \frac{-23}{20}, \ \hat{\tau}_2 = \frac{-7}{20}, \ \hat{\tau}_3 = \frac{13}{20}, \ \hat{\tau}_4 = \frac{17}{20}, \ \hat{\beta}_1 = \frac{35}{20}, \ \hat{\beta}_2 = \frac{-65}{20}, \ \hat{\beta}_3 = \frac{75}{20}, \ \hat{\beta}_4 = \frac{20}{20}, \ \hat{\beta}_5 = \frac{-65}{20}$$

4.23. An industrial engineer is investigating the effect of four assembly methods (A, B, C, D) on the assembly time for a color television component. Four operators are selected for the study. Furthermore, the engineer knows that each assembly method produces such fatigue that the time required for the last assembly may be greater than the time required for the first, regardless of the method. That is, a trend develops in the required assembly time. To account for this source of variability, the engineer uses the Latin square design shown below. Analyze the data from this experiment ($\alpha = 0.05$) and draw appropriate conclusions.

Order of	Operator			
Assembly	1	2	3	4
1	C=10	D=14	A=7	B=8
2	B=7	C=18	D=11	A=8
3	A=5	B=10	C=11	D=9
4	D=10	A=10	B=12	C=14

The MINITAB output below identifies assembly method as having a significant effect on assembly time.

MINITAB Output

```
                         General Linear Model

Factor      Type Levels Values
Order       random    4 1 2 3 4
Operator    random    4 1 2 3 4
Method      fixed     4 A B C D


Analysis of Variance for Time, using Adjusted SS for Tests

Source     DF     Seq SS    Adj SS    Adj MS      F      P
Method      3     72.500    72.500    24.167  13.81  0.004
Order       3     18.500    18.500     6.167   3.52  0.089
Operator    3     51.500    51.500    17.167   9.81  0.010
Error       6     10.500    10.500     1.750
Total      15    153.000
```

4.40. An engineer is studying the mileage performance characteristics of five types of gasoline additives. In the road test he wishes to use cars as blocks; however, because of a time constraint, he must use an incomplete block design. He runs the balanced design with the five blocks that follow. Analyze the data from this experiment (use $\alpha = 0.05$) and draw conclusions.

Additive	Car 1	2	3	4	5
1		17	14	13	12
2	14	14		13	10
3	12		13	12	9
4	13	11	11	12	
5	11	12	10		8

There are several computer software packages that can analyze the incomplete block designs discussed in this chapter. The MINITAB General Linear Model procedure is a widely available package with this capability. The output from this routine for Problem 4.33 follows. The adjusted sums of squares are the appropriate sums of squares to use for testing the difference between the means of the gasoline additives. The gasoline additives have a significant effect on the mileage.

MINITAB Output

```
                            General Linear Model

Factor     Type Levels Values
Additive   fixed      5  1 2 3 4 5
Car        random     5  1 2 3 4 5

Analysis of Variance for Mileage, using Adjusted SS for Tests

Source    DF    Seq SS    Adj SS    Adj MS      F      P
Additive   4   31.7000   35.7333    8.9333   9.81  0.001
Car        4   35.2333   35.2333    8.8083   9.67  0.001
Error     11   10.0167   10.0167    0.9106
Total     19   76.9500
```

4.41. Construct a set of orthogonal contrasts for the data in Problem 4.33. Compute the sum of squares for each contrast.

One possible set of orthogonal contrasts is:

$$H_0 : \mu_4 + \mu_5 = \mu_1 + \mu_2 \qquad (1)$$
$$H_0 : \mu_1 = \mu_2 \qquad (2)$$
$$H_0 : \mu_4 = \mu_5 \qquad (3)$$
$$H_0 : 4\mu_3 = \mu_4 + \mu_5 + \mu_1 + \mu_2 \qquad (4)$$

The sums of squares and F-tests are:

Brand ->	1	2	3	4	5	$\sum c_i Q_i$	SS	F_0
Q_i	33/4	11/4	-3/4	-14/4	-27/4			
(1)	-1	-1	0	1	1	-85/4	30.10	33.06
(2)	1	-1	0	0	0	22/4	4.03	4.426
(3)	0	0	0	-1	1	-13/4	1.41	1.55
(4)	-1	-1	4	-1	-1	-15/4	0.19	0.21

Contrasts (1) and (2) are significant at the 1% and 5% levels, respectively.

4.42. Seven different hardwood concentrations are being studied to determine their effect on the strength of the paper produced. However, the pilot plant can only produce three runs each day. As days may differ, the analyst uses the balanced incomplete block design that follows. Analyze this experiment (use $\alpha = 0.05$) and draw conclusions.

Hardwood Concentration (%)	Days 1	2	3	4	5	6	7
2	114				120		117
4	126	120				119	
6		137	117				134
8	141		129	149			
10		145		150	143		
12			120		118	123	
14				136		130	127

There are several computer software packages that can analyze the incomplete block designs discussed in this chapter. The MINITAB General Linear Model procedure is a widely available package with this capability. The output from this routine for Problem 4.35 follows. The adjusted sums of squares are the appropriate sums of squares to use for testing the difference between the means of the hardwood concentrations.

MINITAB Output

```
                            General Linear Model

Factor     Type Levels Values
Concentr   fixed       7  2 4 6 8 10 12 14
Days       random      7  1 2 3 4 5 6 7

Analysis of Variance for Strength, using Adjusted SS for Tests

Source     DF    Seq SS    Adj SS    Adj MS      F      P
Concentr    6   2037.62   1317.43    219.57   10.42  0.002
Days        6    394.10    394.10     65.68    3.12  0.070
Error       8    168.57    168.57     21.07
Total      20   2600.29
```

4.43. Analyze the data in Example 4.5 using the general regression significance test.

$$
\begin{aligned}
\mu: \quad & 12\hat{\mu} &+3\hat{\tau}_1 &+3\hat{\tau}_2 &+3\hat{\tau}_3 &+3\hat{\tau}_4 &+3\hat{\beta}_1 &+3\hat{\beta}_2 &+3\hat{\beta}_3 &+3\hat{\beta}_4 &= 870 \\
\tau_1: \quad & 3\hat{\mu} &+3\hat{\tau}_1 & & & &+\hat{\beta}_1 &+\hat{\beta}_2 & &+\hat{\beta}_4 &= 218 \\
\tau_2: \quad & 3\hat{\mu} & &+3\hat{\tau}_2 & & & &+\hat{\beta}_2 &+\hat{\beta}_3 &+\hat{\beta}_4 &= 214 \\
\tau_3: \quad & 3\hat{\mu} & & &+3\hat{\tau}_3 & &+\hat{\beta}_1 &+\hat{\beta}_2 &+\hat{\beta}_3 & &= 216 \\
\tau_4: \quad & 3\hat{\mu} & & & &+3\hat{\tau}_4 &+\hat{\beta}_1 & &+\hat{\beta}_3 &+\hat{\beta}_4 &= 222 \\
\beta_1: \quad & 3\hat{\mu} &+\hat{\tau}_1 & &+\hat{\tau}_3 &+\hat{\tau}_4 &+3\hat{\beta}_1 & & & &= 221 \\
\beta_2: \quad & 3\hat{\mu} &+\hat{\tau}_1 &+\hat{\tau}_2 &+\hat{\tau}_3 & & &+3\hat{\beta}_2 & & &= 224 \\
\beta_3: \quad & 3\hat{\mu} & &+\hat{\tau}_2 &+\hat{\tau}_3 &+\hat{\tau}_4 & & &+3\hat{\beta}_3 & &= 207 \\
\beta_4: \quad & 3\hat{\mu} &+\hat{\tau}_1 &+\hat{\tau}_2 & &+\hat{\tau}_4 & & & &+3\hat{\beta}_4 &= 218
\end{aligned}
$$

Applying the constraints $\sum \hat{\tau}_i = \sum \hat{\beta}_j = 0$, we obtain:

$$\hat{\mu} = 870/12, \ \hat{\tau}_1 = -9/8, \ \hat{\tau}_2 = -7/8, \ \hat{\tau}_3 = -4/8, \ \hat{\tau}_4 = 20/8,$$
$$\hat{\beta}_1 = 7/8, \ \hat{\beta}_2 = 24/8, \ \hat{\beta}_3 = -31/8, \ \hat{\beta}_4 = 0/8$$
$$R(\mu, \tau, \beta) = \hat{\mu} y_{..} + \sum_{i=1}^{4} \hat{\tau}_i y_{i.} + \sum_{j=1}^{4} \hat{\beta}_j y_{.j} = 63,152.75$$

with 7 degrees of freedom.

$$\sum \sum y_{ij}^2 = 63,156.00$$
$$SS_E = \sum \sum y_{ij}^2 - R(\mu, \tau, \beta) = 63156.00 - 63152.75 = 3.25.$$

To test $H_o: \tau_i = 0$, the reduced model is $y_{ij} = \mu + \beta_j + \varepsilon_{ij}$. The normal equations used are:

$$
\begin{aligned}
\mu: \quad & 12\hat{\mu} &+3\hat{\beta}_1 &+3\hat{\beta}_2 &+3\hat{\beta}_3 &+3\hat{\beta}_4 &= 870 \\
\beta_1: \quad & 3\hat{\mu} &+3\hat{\beta}_1 & & & &= 221 \\
\beta_2: \quad & 3\hat{\mu} & &+3\hat{\beta}_2 & & &= 224 \\
\beta_3: \quad & 3\hat{\mu} & & &+3\hat{\beta}_3 & &= 207 \\
\beta_4: \quad & 3\hat{\mu} & & & &+3\hat{\beta}_4 &= 218
\end{aligned}
$$

Applying the constraint $\sum \hat{\beta}_j = 0$, we obtain:

$$\hat{\mu} = \frac{870}{12}, \ \hat{\beta}_1 = \frac{7}{6}, \ \hat{\beta}_2 = \frac{13}{6}, \ \hat{\beta}_3 = \frac{-21}{6}, \ \hat{\beta}_4 = \frac{1}{6}$$

$$R(\mu, \beta) = \hat{\mu} y_{..} + \sum_{j=1}^{4} \hat{\beta}_j y_{.j} = 63{,}130.00$$

with 4 degrees of freedom.

$$R(\tau | \mu, \beta) = R(\mu, \tau, \beta) - R(\mu, \beta) = 63152.75 - 63130.00 = 22.75 = SS_{Treatments}$$

with $7 - 4 = 3$ degrees of freedom. $R(\tau | \mu, \beta)$ is used to test $H_o: \tau_i = 0$.

The sum of squares for blocks is found from the reduced model $y_{ij} = \mu + \tau_i + \varepsilon_{ij}$. The normal equations used are:

Model Restricted to $\beta_j = 0$:

$$
\begin{array}{llllll}
\mu: & 12\hat{\mu} & +3\hat{\tau}_1 & +3\hat{\tau}_2 & +3\hat{\tau}_3 & +3\hat{\tau}_4 & = 870 \\
\tau_1: & 3\hat{\mu} & +3\hat{\tau}_1 & & & & = 218 \\
\tau_2: & 3\hat{\mu} & & +3\hat{\tau}_2 & & & = 214 \\
\tau_3: & 3\hat{\mu} & & & +3\hat{\tau}_3 & & = 216 \\
\tau_4: & 3\hat{\mu} & & & & +3\hat{\tau}_4 & = 222
\end{array}
$$

The sum of squares for blocks is found as in Example 4.5. We may use the method shown above to find an adjusted sum of squares for blocks from the reduced model, $y_{ij} = \mu + \tau_i + \varepsilon_{ij}$.

4.44. Prove that $\dfrac{k \sum_{i=1}^{a} Q_i^2}{(\lambda a)}$ is the adjusted sum of squares for treatments in a BIBD.

We may use the general regression significance test to derive the computational formula for the adjusted treatment sum of squares. We will need the following:

$$\hat{\tau}_i = \frac{kQ_i}{(\lambda a)}, \quad kQ_i = ky_{i.} - \sum_{i=1}^{b} n_{ij} y_{.j}$$

$$R(\mu, \tau, \beta) = \hat{\mu} y_{..} + \sum_{i=1}^{a} \hat{\tau}_i y_{i.} + \sum_{j=1}^{b} \hat{\beta}_j y_{.j}$$

and the sum of squares we need is:

$$R(\tau | \mu, \beta) = \hat{\mu} y_{..} + \sum_{i=1}^{a} \hat{\tau}_i y_{i.} + \sum_{j=1}^{b} \hat{\beta}_j y_{.j} - \sum_{j=1}^{b} \frac{y_{.j}^2}{k}$$

The normal equation for β is, from Equation 4-35,

$$\beta : k\hat{\mu} + \sum_{i=1}^{a} n_{ij}\hat{\tau}_i + k\hat{\beta}_j = y_{.j}$$

and from this we have:

$$ky_{.j}\hat{\beta}_j = y_{.j}^2 - ky_{.j}\hat{\mu} - y_{.j}\sum_{i=1}^{a} n_{ij}\hat{\tau}_i$$

therefore,

$$R(\tau|\mu,\beta) = \hat{\mu}y_{..} + \sum_{i=1}^{a}\hat{\tau}_i y_{i.} + \sum_{j=1}^{b}\left[\frac{y_{.j}^2}{k} - \frac{k\hat{\mu}y_{.j}}{k} - \frac{y_{.j}\sum_{i=1}^{a} n_{ij}\hat{\tau}_i}{k} - \frac{y_{.j}^2}{k}\right]$$

$$R(\tau|\mu,\beta) = \sum_{i=1}^{a}\hat{\tau}_i\left(y_{i.} - \frac{1}{k}\sum_{i=1}^{a} n_{ij}y_{.j}\right) = \sum_{i=1}^{a} Q_i\left(\frac{kQ_i}{\lambda a}\right) = k\sum_{i=1}^{a}\left(\frac{Q_i^2}{\lambda a}\right) \equiv SS_{Treatments(adjusted)}$$

4.45. An experimenter wishes to compare four treatments in blocks of two runs. Find a BIBD for this experiment with six blocks.

Treatment	Block 1	Block 2	Block 3	Block 4	Block 5	Block 6
1	X	X	X			
2	X			X	X	
3		X		X		X
4			X		X	X

Note that the design is formed by taking all combinations of the 4 treatments 2 at a time. The parameters of the design are $\lambda = 1$, $a = 4$, $b = 6$, $k = 3$, and $r = 2$.

4.49. Verify that a BIBD with the parameters $a = 8$, $r = 8$, $k = 4$, and $b = 16$ does not exist.

These conditions imply that $\lambda = \dfrac{r(k-1)}{a-1} = \dfrac{8(3)}{7} = \dfrac{24}{7}$, which is not an integer, so a balanced design with these parameters cannot exist.

4.50. Show that the variance of the intrablock estimators $\{\hat{\tau}_i\}$ is $\dfrac{k((a-1))\sigma^2}{(\lambda a^2)}$.

Note that $\hat{\tau}_i = \dfrac{kQ_i}{(\lambda a)}$, and $Q_i = y_{i.} - \dfrac{1}{k}\sum\limits_{j=1}^{b} n_{ij} y_{.j}$, and $kQ_i = ky_{i.} - \sum\limits_{j=1}^{b} n_{ij} y_{.j} = (k-1)y_{i.} - \left(\sum\limits_{j=1}^{b} n_{ij} y_{.j} - y_{i.}\right)$

$y_{i.}$ contains r observations, and the quantity in the parentheses is the sum of $r(k-1)$ observations, not including treatment i. Therefore,

$$V(kQ_i) = k^2 V(Q_i) = r(k-1)^2 \sigma^2 + r(k-1)\sigma^2$$

or

$$V(Q_i) = \frac{1}{k^2}\left[r(k-1)\sigma^2\{(k-1)+1\}\right] = \frac{r(k-1)\sigma^2}{k}$$

To find $V(\hat{\tau}_i)$, note that:

$$V(\hat{\tau}_i) = \left(\frac{k}{\lambda a}\right)^2 V(Q)_i = \left(\frac{k}{\lambda a}\right)^2 \frac{r(k-1)}{k}\sigma^2 = \frac{kr(k-1)}{(\lambda a)^2}\sigma^2$$

However, since $\lambda(a-1) = r(k-1)$, we have:

$$V(\hat{\tau}_i) = \frac{k(a-1)}{\lambda a^2}\sigma^2$$

Furthermore, the $\{\hat{\tau}_i\}$ are not independent, this is required to show that $V(\hat{\tau}_i - \hat{\tau}_j) = \dfrac{2k}{\lambda a}\sigma^2$

CHAPTER 5

Introduction to Factorial Designs

LEARNING OBJECTIVES

After completing this chapter, you will be able to:

1. Design, conduct, and analyze experiments involving a two-factor factorial.

2. Interpret main effects and interactions.

3. Use residual analysis to investigate the adequacy of the factorial design model and check the validity of the underlying assumptions.

4. Understand how the factorial design concept generalizes to more than two factors.

5. Understand how the blocking principle extends to a factorial design.

KEY CONCEPTS AND IDEAS

1. Main effect

2. Interaction

3. Statistical model for a factorial experiment

4. Analysis of variance for a factorial experiment

5. Quantitative and qualitative factors

6. Response curves and surfaces

7. Blocking

Exercises

5.6. An article in *Industrial Quality Control* (1956, pp. 5-8) describes an experiment to investigate the effect of the type of glass and the type of phosphor on the brightness of a television tube. The response variable is the current necessary (in microamps) to obtain a specified brightness level. The data are as follows:

Glass	Phosphor Type		
Type	1	2	3
	280	300	290
1	290	310	285
	285	295	290
	230	260	220
2	235	240	225
	240	235	230

(a) Is there any indication that either factor influences brightness? Use $\alpha = 0.05$.

Both factors, phosphor type (A) and glass type (B), influence brightness.

Design-Expert Output

Response: Current in microamps
ANOVA for Selected Factorial Model
Analysis of variance table [Partial sum of squares]

Source	Sum of Squares	DF	Mean Square	F Value	Prob > F	
Model	15516.67	5	3103.33	58.80	< 0.0001	significant
A	933.33	2	466.67	8.84	0.0044	
B	14450.00	1	14450.00	273.79	< 0.0001	
AB	133.33	2	66.67	1.26	0.3178	
Residual	633.33	12	52.78			
Lack of Fit	0.000	0				
Pure Error	633.33	12	52.78			
Cor Total	16150.00	17				

The Model F-value of 58.80 implies the model is significant. There is only a 0.01% chance that a "Model F-Value" this large could occur due to noise.

Values of "Prob > F" less than 0.0500 indicate model terms are significant. In this case A, B are significant model terms.

(b) Do the two factors interact? Use $\alpha = 0.05$.

There is no interaction effect.

(c) Analyze the residuals from this experiment.

The residual plot of residuals versus phosphor content indicates a very slight inequality of variance. It is not serious enough to be of concern, however.

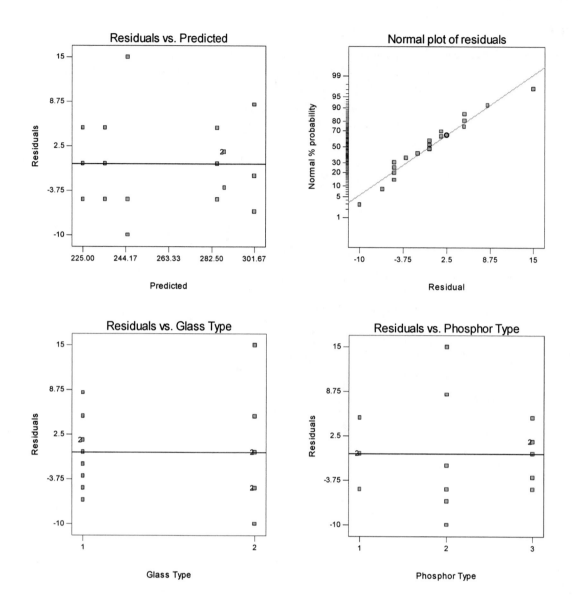

5.7. Johnson and Leone (*Statistics and Experimental Design in Engineering and the Physical Sciences*, Wiley 1977) describe an experiment to investigate the warping of copper plates. The two factors studied were the temperature and the copper content of the plates. The response variable was a measure of the amount of warping. The data were as follows:

Temperature (°C)	Copper	Content (%)		
	40	60	80	100
50	17,20	16,21	24,22	28,27
75	12,9	18,13	17,12	27,31
100	16,12	18,21	25,23	30,23
125	21,17	23,21	23,22	29,31

(a) Is there any indication that either factor affects the amount of warping? Is there any interaction between the factors? Use $\alpha = 0.05$.

Both factors, copper content (A) and temperature (B), affect warping, while the interaction does not.

Design-Expert Output

Response: Warping
    ANOVA for Selected Factorial Model
Analysis of variance table [Partial sum of squares]

Source	Sum of Squares	DF	Mean Square	F Value	Prob > F	
Model	968.22	15	64.55	9.52	< 0.0001	significant
A	698.34	3	232.78	34.33	< 0.0001	
B	156.09	3	52.03	7.67	0.0021	
AB	113.78	9	12.64	1.86	0.1327	
Residual	108.50	16	6.78			
Lack of Fit	0.000	0				
Pure Error	108.50	16	6.78			
Cor Total	1076.72	31				

The Model F-value of 9.52 implies the model is significant. There is only
a 0.01% chance that a "Model F-Value" this large could occur due to noise.

Values of "Prob > F" less than 0.0500 indicate model terms are significant.
In this case A, B are significant model terms.

(b) Analyze the residuals from this experiment.

There is nothing unusual about the residual plots.

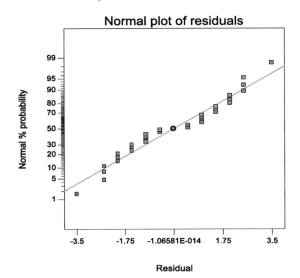

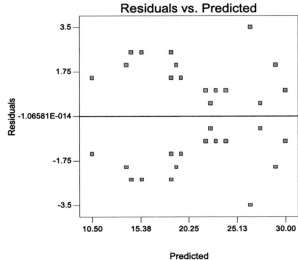

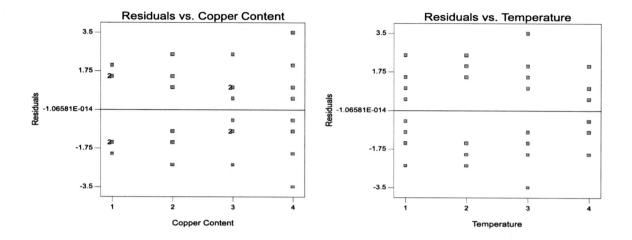

(c) Plot the average warping at each level of copper content and compare them to an appropriately scaled t distribution. Describe the differences in the effects of the different levels of copper content on warping. If low warping is desirable, what level of copper content would you specify?

Design-Expert Output

Factor	Name	Level	Low Level	High Level				
A	Copper Content	40	40	100				
B	Temperature	Average	50	125				
	Prediction	**SE Mean**	**95% CI low**	**95% CI high**	**SE Pred**	**95% PI low**	**95% PI high**	
Warping	15.5	0.92	13.55	17.45	2.76	9.64	21.36	

Factor	Name	Level	Low Level	High Level				
A	Copper Content	60	40	100				
B	Temperature	Average	50	125				
	Prediction	**SE Mean**	**95% CI low**	**95% CI high**	**SE Pred**	**95% PI low**	**95% PI high**	
Warping	18.875	0.92	16.92	20.83	2.76	13.02	24.73	

Factor	Name	Level	Low Level	High Level				
A	Copper Content	80	40	100				
B	Temperature	Average	50	125				
	Prediction	**SE Mean**	**95% CI low**	**95% CI high**	**SE Pred**	**95% PI low**	**95% PI high**	
Warping	21	0.92	19.05	22.95	2.76	15.14	26.86	

Factor	Name	Level	Low Level	High Level				
A	Copper Content	100	40	100				
B	Temperature	Average	50	125				
	Prediction	**SE Mean**	**95% CI low**	**95% CI high**	**SE Pred**	**95% PI low**	**95% PI high**	
Warping	28.25	0.92	26.30	30.20	2.76	22.39	34.11	

Use a copper content of 40 for the lowest warping.

$$S = \sqrt{\frac{MS_E}{b}} = \sqrt{\frac{6.78125}{8}} = 0.92$$

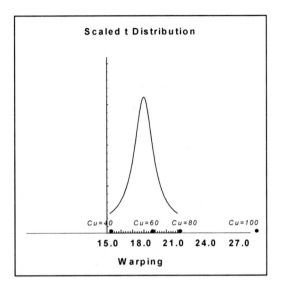

(d) Suppose that temperature cannot be easily controlled in the environment in which the copper plates are to be used. Does this change your answer for part (c)?

Use a copper of content 40. This is the same as for part (c).

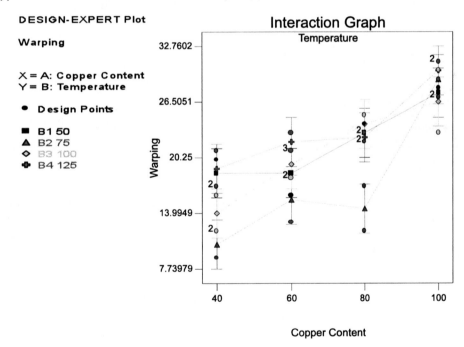

5.8. The factors that influence the breaking strength of a synthetic fiber are being studied. Four production machines and three operators are chosen and a factorial experiment is run using fiber from the same production batch. The results are as follows:

Operator	Machine			
	1	2	3	4
1	109	110	108	110
	110	115	109	108
2	110	110	111	114
	112	111	109	112
3	116	112	114	120
	114	115	119	117

(a) Analyze the data and draw conclusions. Use $\alpha = 0.05$.

Only the Operator (A) effect is significant.

Design-Expert Output

Response: Stength
ANOVA for Selected Factorial Model
Analysis of variance table [Partial sum of squares]

Source	Sum of Squares	DF	Mean Square	F Value	Prob > F	
Model	217.46	11	19.77	5.21	0.0041	significant
A	160.33	2	80.17	21.14	0.0001	
B	12.46	3	4.15	1.10	0.3888	
AB	44.67	6	7.44	1.96	0.1507	
Residual	45.50	12	3.79			
Lack of Fit	0.000	0				
Pure Error	45.50	12	3.79			
Cor Total	262.96	23				

The Model F-value of 5.21 implies the model is significant.
There is only a 0.41% chance that a "Model F-Value" this large could occur due to noise.

Values of "Prob > F" less than 0.0500 indicate model terms aresignificant.
In this case A are significant model terms.

(b) Prepare appropriate residual plots and comment on the model's adequacy.

The plot of residuals versus predicted shows that variance increases very slightly with strength. There is no indication of a severe problem.

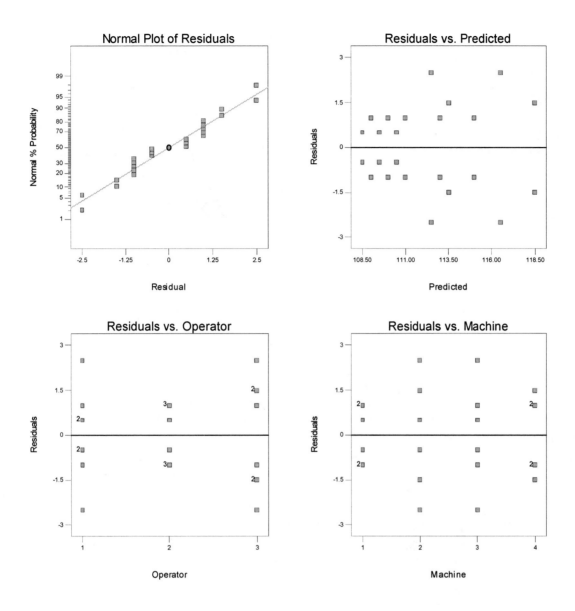

5.17. Consider the three-factor model

$$y_{ijk} = \mu + \tau_i + \beta_j + \gamma_k + (\tau\beta)_{ij} + (\beta\gamma)_{jk} + \varepsilon_{ijk} \quad \begin{cases} i = 1,2,...,a \\ j = 1,2,...,b \\ k = 1,2,...,c \end{cases}$$

Notice that there is only one replicate. Assuming all the factors are fixed, write down the analysis of variance table, including the expected mean squares. What would you use as the "experimental error" to test hypotheses?

Source	Degrees of Freedom	Expected Mean Square
A	$a{-}1$	$\sigma^2 + bc\sum_{i=1}^{a}\dfrac{\tau_i^2}{(a-1)}$
B	$b{-}1$	$\sigma^2 + ac\sum_{j=1}^{b}\dfrac{\beta_j^2}{(b-1)}$
C	$c{-}1$	$\sigma^2 + ab\sum_{k=1}^{c}\dfrac{\gamma_k^2}{(c-1)}$
AB	$(a{-}1)(b{-}1)$	$\sigma^2 + c\sum_{i=1}^{a}\sum_{j=1}^{b}\dfrac{(\tau\beta)_{ij}^2}{(a-1)(b-1)}$
BC	$(b{-}1)(c{-}1)$	$\sigma^2 + a\sum_{j=1}^{b}\sum_{k=1}^{c}\dfrac{(\beta\gamma)_{jk}^2}{(b-1)(c-1)}$
Error (AC + ABC)	$b(a{-}1)(c{-}1)$	σ^2
Total	$abc{-}1$	

5.19. The quality control department of a fabric finishing plant is studying the effect of several factors on the dyeing of cotton-synthetic cloth used to manufacture men's shirts. Three operators, three cycle times, and two temperatures were selected, and three small specimens of cloth were dyed under each set of conditions. The finished cloth was compared to a standard, and a numerical score was assigned. The results are as follows. Analyze the data and draw conclusions. Comment on the model's adequacy.

	Temperature						
	300°				350°		
	Operator				Operator		
Cycle Time	1	2	3		1	2	3
40	23	27	31		24	38	34
	24	28	32		23	36	36
	25	26	29		28	35	39
50	36	34	33		37	34	34
	35	38	34		39	38	36
	36	39	35		35	36	31
60	28	35	26		26	36	28
	24	35	27		29	37	26
	27	34	25		25	34	24

All three main effects, and the *AB, AC*, and *ABC* interactions are significant. There is nothing unusual about the residual plots.

Design-Expert Output

Response: Score
ANOVA for Selected Factorial Model
Analysis of variance table [Partial sum of squares]

Source	Sum of Squares	DF	Mean Square	F Value	Prob > F	
Model	1239.33	17	72.90	22.24	< 0.0001	significant
A	*436.00*	*2*	*218.00*	*66.51*	*< 0.0001*	
B	*261.33*	*2*	*130.67*	*39.86*	*< 0.0001*	
C	*50.07*	*1*	*50.07*	*15.28*	*0.0004*	
AB	*355.67*	*4*	*88.92*	*27.13*	*< 0.0001*	
AC	*78.81*	*2*	*39.41*	*12.02*	*0.0001*	
BC	*11.26*	*2*	*5.63*	*1.72*	*0.1939*	
ABC	*46.19*	*4*	*11.55*	*3.52*	*0.0159*	
Residual	118.00	36	3.28			
Lack of Fit	*0.000*	*0*				
Pure Error	*118.00*	*36*	*3.28*			
Cor Total	1357.33	53				

The Model F-value of 22.24 implies the model is significant. There is only
a 0.01% chance that a "Model F-Value" this large could occur due to noise.

Values of "Prob > F" less than 0.0500 indicate model terms are significant.
In this case A, B, C, AB, AC, ABC are significant model terms.

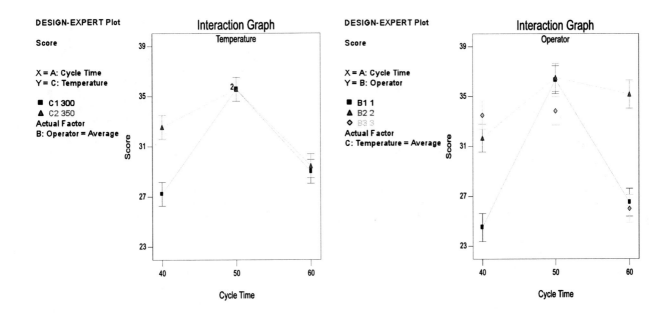

DESIGN-EXPERT Plot

Score

X = A: Cycle Time
Y = C: Temperature

■ C1 300
▲ C2 350
Actual Factor
B: Operator = Average

DESIGN-EXPERT Plot

Score

X = A: Cycle Time
Y = B: Operator

■ B1 1
▲ B2 2
◇ B3 3
Actual Factor
C: Temperature = Average

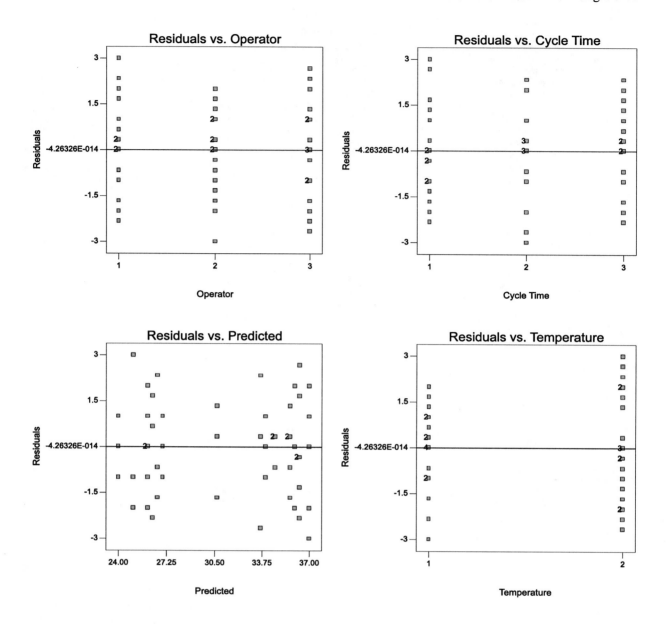

5.23. Consider the data in Problem 5.8. Analyze the data, assuming that replicates are blocks.

Design-Expert Output

Response: Strength						
ANOVA for Selected Factorial Model						
Analysis of variance table [Partial sum of squares]						
Source	Sum of Squares	DF	Mean Square	F Value	Prob > F	
Block	2.04	1	2.04			
Model	217.46	11	19.77	5.00	0.0064	significant
A	160.33	2	80.17	20.29	0.0002	
B	12.46	3	4.15	1.05	0.4087	
AB	44.67	6	7.44	1.88	0.1716	
Residual	43.46	11	3.95			
Cor Total	262.96	23				

> The Model F-value of 5.00 implies the model is significant. There is only
> a 0.64% chance that a "Model F-Value" this large could occur due to noise.
>
> Values of "Prob > F" less than 0.0500 indicate model terms are significant.
> In this case A are significant model terms.

Only the operator factor (A) is significant. This agrees with the analysis in Problem 5.8.

5.24. An article in the *Journal of Testing and Evaluation* (Vol. 16, no.2, pp. 508-515) investigated the effects of cyclic loading and environmental conditions on fatigue crack growth at a constant 22 MPa stress for a particular material. The data from this experiment are shown below (the response is crack growth rate):

		Environment	
Frequency	Air	H_2O	Salt H_2O
	2.29	2.06	1.90
10	2.47	2.05	1.93
	2.48	2.23	1.75
	2.12	2.03	2.06
	2.65	3.20	3.10
1	2.68	3.18	3.24
	2.06	3.96	3.98
	2.38	3.64	3.24
	2.24	11.00	9.96
0.1	2.71	11.00	10.01
	2.81	9.06	9.36
	2.08	11.30	10.40

(a) Analyze the data from this experiment (use $\alpha = 0.05$).

Design-Expert Output

Response: Crack Growth
ANOVA for Selected Factorial Model
Analysis of variance table [Partial sum of squares]

Source	Sum of Squares	DF	Mean Square	F Value	Prob > F	
Model	376.11	8	47.01	234.02	< 0.0001	significant
A	209.89	2	104.95	522.40	< 0.0001	
B	64.25	2	32.13	159.92	< 0.0001	
AB	101.97	4	25.49	126.89	< 0.0001	
Residual	5.42	27	0.20			
Lack of Fit	0.000	0				
Pure Error	5.42	27	0.20			
Cor Total	381.53	35				

The Model F-value of 234.02 implies the model is significant. There is only
a 0.01% chance that a "Model F-Value" this large could occur due to noise.

Values of "Prob > F" less than 0.0500 indicate model terms are significant.
In this case A, B, AB are significant model terms.

Both frequency and environment, as well as their interaction, are significant.

(b) Analyze the residuals.

The residual plots indicate that there may be some problem with inequality of variance as well as the possibility of an outlier. This is particularly noticeable on the plot of residuals versus predicted response, normal probability plot of residuals, and the plot of residuals versus frequency.

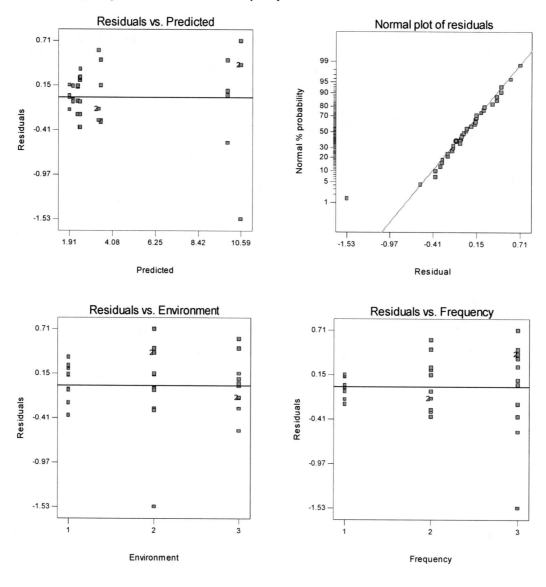

(c) Repeat the analyses from parts (a) and (b) using ln(*y*) as the response. Comment on the results.

Design-Expert Output

Response: Crack Growth			**Transform: Natural log**		**Constant: 0.000**	
ANOVA for Selected Factorial Model						
Analysis of variance table [Partial sum of squares]						
	Sum of		**Mean**	**F**		
Source	**Squares**	**DF**	**Square**	**Value**	**Prob > F**	
Model	13.46	8	1.68	179.57	< 0.0001	significant
A	*7.57*	*2*	*3.79*	*404.09*	*< 0.0001*	
B	*2.36*	*2*	*1.18*	*125.85*	*< 0.0001*	
AB	*3.53*	*4*	*0.88*	*94.17*	*< 0.0001*	
Residual	0.25	27	9.367E-003			
Lack of Fit	*0.000*	*0*				
Pure Error	0.25	27	9.367E-003			
Cor Total	13.71	35				

The Model F-value of 179.57 implies the model is significant. There is only a 0.01% chance that a "Model F-Value" this large could occur due to noise.

Values of "Prob > F" less than 0.0500 indicate model terms are significant. In this case A, B, AB are significant model terms.

Both frequency and environment, as well as their interaction, are significant. The residual plots from the analysis of the transformed data look better; there is no indication of problems with the assumptions or model adequacy.

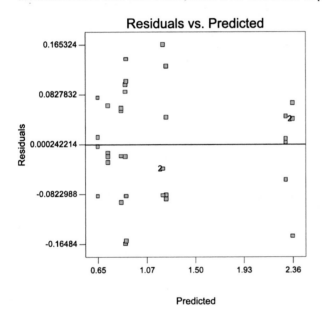

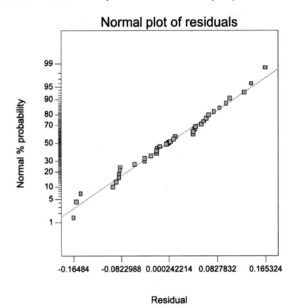

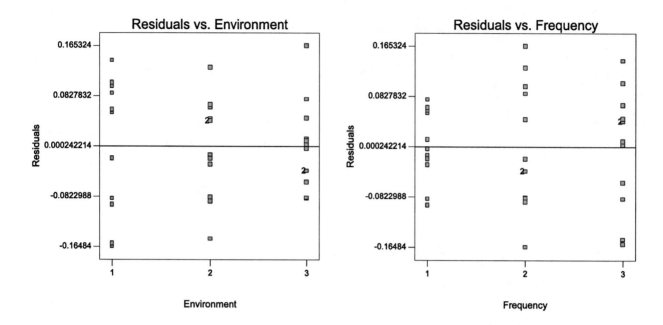

CHAPTER **6**

The 2^k Factorial Design

LEARNING OBJECTIVES

After completing this chapter, you will be able to:

1. Plan, conduct, and analyze experiments involving a 2^k factorial design, including both replicated and unreplicated versions of the design.

2. Fit a model to the data from a 2^k factorial design and interpret this model through interaction graphs, contour plots, and response surface plots.

3. Use residual analysis to investigate the adequacy of the model and check the validity of the underlying assumptions.

4. Use center points in a 2^k factorial design to investigate the necessity of adding pure quadratic terms to the model.

KEY CONCEPTS AND IDEAS

1. Factor screening experiment
2. 2^k factorial design
3. Main effect
4. Interaction
5. Statistical model for the data
6. Orthogonal design

7. Normal probability plot of effects
8. Design projection
9. Response surface
10. Contour plot
11. Center points
12. Second-order model

Exercises

6.1. An engineer is interested in the effects of cutting speed (A), tool geometry (B), and cutting angle on the life (in hours) of a machine tool. Two levels of each factor are chosen, and three replicates of a 2^3 factorial design are run. The results are as follows:

A	B	C	Treatment Combination	Replicate I	Replicate II	Replicate III
-	-	-	(1)	22	31	25
+	-	-	a	32	43	29
-	+	-	b	35	34	50
+	+	-	ab	55	47	46
-	-	+	c	44	45	38
+	-	+	ac	40	37	36
-	+	+	bc	60	50	54
+	+	+	abc	39	41	47

(a) Estimate the factor effects. Which effects appear to be large?

From the normal probability plot of effects below, factors B, C, and the AC interaction appear to be significant.

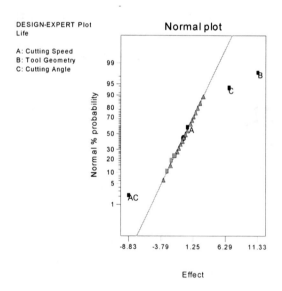

(b) Use the analysis of variance to confirm your conclusions for part (a).

The analysis of variance confirms the significance of factors B, C, and the AC interaction.

Design-Expert Output

Response: Life in hours						
ANOVA for Selected Factorial Model						
Analysis of variance table [Partial sum of squares]						
Source	Sum of Squares	DF	Mean Square	F Value	Prob > F	
Model	1612.67	7	230.38	7.64	0.0004	significant
A	0.67	1	0.67	0.022	0.8837	
B	770.67	1	770.67	25.55	0.0001	
C	280.17	1	280.17	9.29	0.0077	
AB	16.67	1	16.67	0.55	0.4681	
AC	468.17	1	468.17	15.52	0.0012	
BC	48.17	1	48.17	1.60	0.2245	
ABC	28.17	1	28.17	0.93	0.3483	
Pure Error	482.67	16	30.17			
Cor Total	2095.33	23				

The Model F-value of 7.64 implies the model is significant. There is only
a 0.04% chance that a "Model F-Value" this large could occur due to noise.

The reduced model ANOVA is shown below. Factor A was included to maintain hierarchy.

Design-Expert Output

Response: Life in hours						
ANOVA for Selected Factorial Model						
Analysis of variance table [Partial sum of squares]						
Source	Sum of Squares	DF	Mean Square	F Value	Prob > F	
Model	1519.67	4	379.92	12.54	< 0.0001	significant
A	0.67	1	0.67	0.022	0.8836	
B	770.67	1	770.67	25.44	< 0.0001	
C	280.17	1	280.17	9.25	0.0067	
AC	468.17	1	468.17	15.45	0.0009	
Residual	575.67	19	30.30			
Lack of Fit	93.00	3	31.00	1.03	0.4067	not significant
Pure Error	482.67	16	30.17			
Cor Total	2095.33	23				

The Model F-value of 12.54 implies the model is significant. There is only
a 0.01% chance that a "Model F-Value" this large could occur due to noise.

Effects B, C and AC are significant at 1%.

(c) Write down a regression model for predicting tool life (in hours) based on the results of this experiment.

$$y_{ijk} = 40.8333 + 0.1667x_A + 5.6667x_B + 3.4167x_C + 4.4167x_A x_C$$

Design-Expert Output

Factor	Coefficient Estimate	DF	Standard Error	95% CI Low	95% CI High	VIF
Intercept	40.83	1	1.12	38.48	43.19	
A-Cutting Speed	0.17	1	1.12	-2.19	2.52	1.00
B-Tool Geometry	5.67	1	1.12	3.31	8.02	1.00
C-Cutting Angle	3.42	1	1.12	1.06	5.77	1.00
AC	-4.42	1	1.12	-6.77	-2.06	1.00

Final Equation in Terms of Coded Factors:

Life	=	
+40.83		
+0.17	* A	
+5.67	* B	
+3.42	* C	
-4.42	* A * C	

Final Equation in Terms of Actual Factors:

Life	=	
+40.83333		
+0.16667	* Cutting Speed	
+5.66667	* Tool Geometry	
+3.41667	* Cutting Angle	
-4.41667	* Cutting Speed * Cutting Angle	

The equation in part (c) and shown in the computer output form satisfies a hierarchial model preference. For example, a hierarchial model with a significant interaction would also include all of the main effects referenced in the interaction even if they are not significant.

(d) Analyze the residuals. Are there any obvious problems?

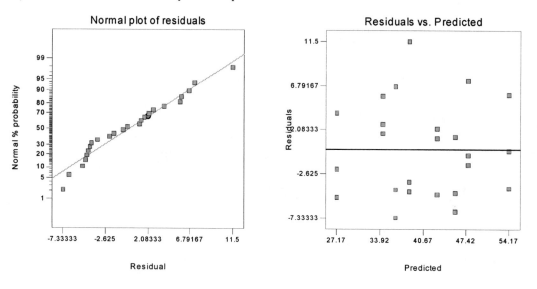

There is nothing unusual about the residual plots.

(e) On the basis of an analysis of main effects and interaction plots, what coded factor levels of A, B, and C would you recommend using?

Since B has a positive effect, set B at the high level to increase life. The AC interaction plot reveals that life would be maximized with C at the high level and A at the low level.

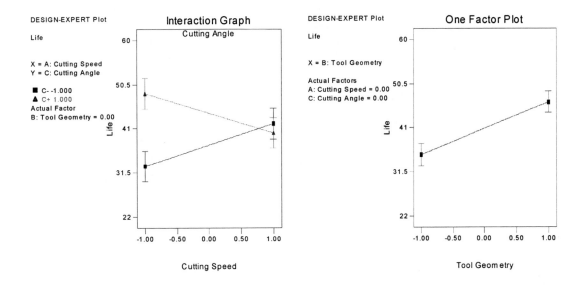

6.2. Reconsider part (c) of Problem 6.1. Use the regression model to generate response surface and contour plots of the tool life response. Interpret these plots. Do they provide insight regarding the desirable operating conditions for this process?

The response surface plot and the contour plot in terms of factors *A* and *C* with *B* at the high level are shown below. They show the curvature due to the *AC* interaction. These plots make it easy to see the region of greatest tool life.

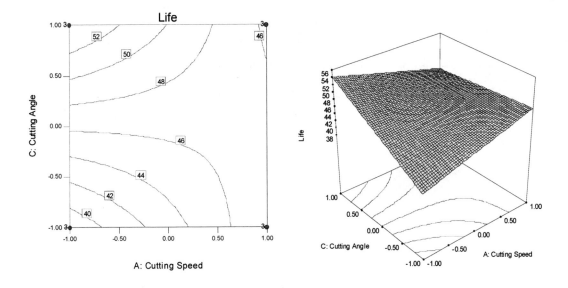

6.5. A router is used to cut locating notches on a printed circuit board. The vibration level at the surface of the board as it is cut is considered to be a major source of dimensional variation in the notches. Two factors are thought to influence vibration: bit size (A) and cutting speed (B). Two bit sizes (1/16 and 1/8 inch) and two speeds (40 and 90 rpm) are selected, and four boards are cut at each set of conditions shown below. The response variable is vibration measured as a resultant vector of three accelerometers (x, y, and z) on each test circuit board.

A	B	Treatment Combination	I	II	III	IV
-	-	(1)	18.2	18.9	12.9	14.4
+	-	a	27.2	24.0	22.4	22.5
-	+	b	15.9	14.5	15.1	14.2
+	+	ab	41.0	43.9	36.3	39.9

(a) Analyze the data from this experiment.

Design-Expert Output

Response: Vibration
ANOVA for Selected Factorial Model
Analysis of variance table [Partial sum of squares]

Source	Sum of Squares	DF	Mean Square	F Value	Prob > F	
Model	1638.11	3	546.04	91.36	< 0.0001	significant
A	1107.23	1	1107.23	185.25	< 0.0001	
B	227.26	1	227.26	38.02	< 0.0001	
AB	303.63	1	303.63	50.80	< 0.0001	
Residual	71.72	12	5.98			
Lack of Fit	0.000	0				
Pure Error	71.72	12	5.98			
Cor Total	1709.83	15				

The Model F-value of 91.36 implies the model is significant. There is only a 0.01% chance that a "Model F-Value" this large could occur due to noise.

(b) Construct a normal probability plot of the residuals, and plot the residuals versus the predicted vibration level. Interpret these plots.

There is nothing unusual about the residual plots shown below.

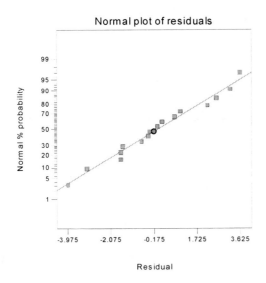

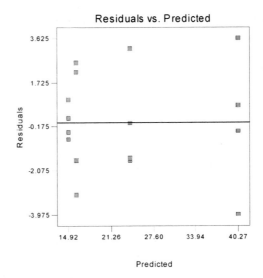

(c) Draw the *AB* interaction plot. Interpret this plot. What levels of bit size and speed would you recommend for routine operation?

To reduce the vibration, use the smaller bit. Once the small bit is specified, either speed will work equally well, because the slope of the curve relating vibration to speed for the small tip is approximately zero. The process is robust to speed changes if the small bit is used.

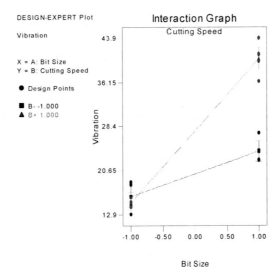

6.6. Reconsider the experiment described in Problem 6.1. Suppose that the experimenter only performed the eight trials from replicate I. In addition, he ran four center points and obtained the following response values: 36, 40, 43, 45.

(a) Estimate the factor effects. Which effects are large?

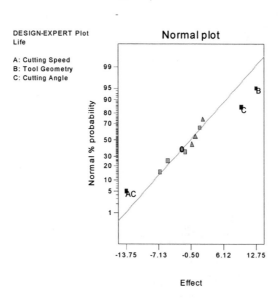

Effects *B, C,* and *AC* appear to be large.

(b) Perform an analysis of variance, including a check for pure quadratic curvature. What are your conclusions?

$$SS_{PureQuadratic} = \frac{n_F n_C (\bar{y}_F - \bar{y}_C)^2}{n_F + n_C} = \frac{(8)(4)(40.875 - 41.000)^2}{8 + 4} = 0.0417$$

Design-Expert Output

Response: Life in hours
    ANOVA for Selected Factorial Model
Analysis of variance table [Partial sum of squares]

Source	Sum of Squares	DF	Mean Square	F Value	Prob > F	
Model	1048.88	7	149.84	9.77	0.0439	significant
A	3.13	1	3.13	0.20	0.6823	
B	325.13	1	325.13	21.20	0.0193	
C	190.12	1	190.12	12.40	0.0389	
AB	6.13	1	6.13	0.40	0.5722	
AC	378.12	1	378.12	24.66	0.0157	
BC	55.12	1	55.12	3.60	0.1542	
ABC	91.12	1	91.12	5.94	0.0927	
Curvature	0.042	1	0.042	2.717E-003	0.9617	not significant
Pure Error	46.00	3	15.33			
Cor Total	1094.92	11				

The Model F-value of 9.77 implies the model is significant. There is only
a 4.39% chance that a "Model F-Value" this large could occur due to noise.

The "Curvature F-value" of 0.00 implies the curvature (as measured by difference between the
average of the center points and the average of the factorial points) in the design space is not
significant relative to the noise. There is a 96.17% chance that a "Curvature F-value"
this large could occur due to noise.

Design-Expert Output

Response: Life in hours
ANOVA for Selected Factorial Model
Analysis of variance table [Partial sum of squares]

Source	Sum of Squares	DF	Mean Square	F Value	Prob > F	
Model	896.50	4	224.13	7.91	0.0098	significant
A	3.13	1	3.13	0.11	0.7496	
B	325.12	1	325.12	11.47	0.0117	
C	190.12	1	190.12	6.71	0.0360	
AC	378.12	1	378.12	13.34	0.0082	
Residual	198.42	7	28.35			
Lack of Fit	152.42	4	38.10	2.49	0.2402	not significant
Pure Error	46.00	3	15.33			
Cor Total	1094.92	11				

The Model F-value of 7.91 implies the model is significant. There is only
a 0.98% chance that a "Model F-Value" this large could occur due to noise.

Effects B, C and AC are significant at 5%. There is no effect of curvature.

(c) Write down an appropriate model for predicting tool life, based on the results of this experiment. Does this
   model differ in any substantial way from the model in Problem 6.1, part (c)?

The model shown in the Design-Expert output below does not differ substantially from the model in Problem 6.1,
part (c).

Design-Expert Output

Final Equation in Terms of Coded Factors:

Life	=	
+40.88		
+0.62	* A	
+6.37	* B	
+4.87	* C	
-6.88	* A * C	

(d) Analyze the residuals.

There is nothing unusual about the residual plots shown below.

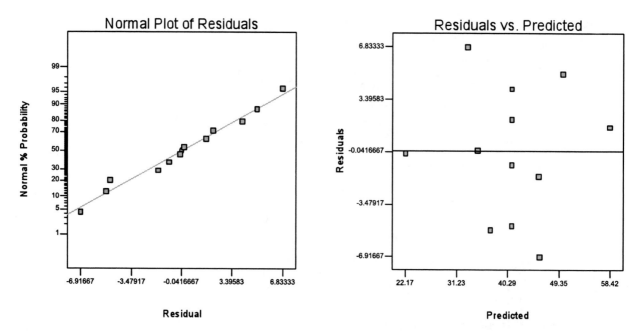

(e) What conclusions would you draw about the appropriate operating conditions for this process?

To maximize life, run with B at the high level, A at the low level, and C at the high level.

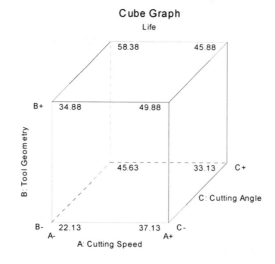

6.20. Consider a variation of the bottle filling experiment from Example 5.3. Suppose that only two levels of carbonation are used so that the experiment is a 2^3 factorial design with two replicates. The data are shown in Table P6.3.

Table P6.3

Run	Coded Factors			Fill Height Deviation	
	A	B	C	Replicate 1	Replicate 2
1	-	-	-	-3	-1
2	+	-	-	0	1
3	-	+	-	-1	0
4	+	+	-	2	3
5	-	-	+	-1	0
6	+	-	+	2	1
7	-	+	+	1	1
8	+	+	+	6	5

	Factor Levels	
	Low (-1)	High (+1)
A (%)	10	12
B (psi)	25	30
C (b/m)	200	250

(a) Analyze the data from this experiment. Which factors significantly affect fill height deviation?

The half normal probability plot of effects shown below identifies the factors A, B, and C as being significant and the AB interaction as being marginally significant. The analysis of variance in the Design-Expert output below confirms that factors A, B, and C are significant and the AB interaction is marginally significant.

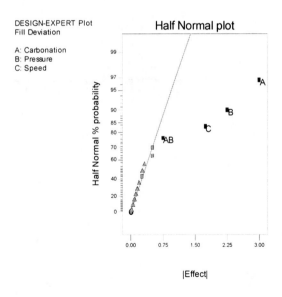

Design-Expert Output

Response: Fill Deviation						
ANOVA for Selected Factorial Model						
Analysis of variance table [Partial sum of squares]						
	Sum of		**Mean**	**F**		
Source	**Squares**	**DF**	**Square**	**Value**	**Prob > F**	
Model	70.75	4	17.69	26.84	< 0.0001	significant
A	*36.00*	*1*	*36.00*	*54.62*	*< 0.0001*	
B	*20.25*	*1*	*20.25*	*30.72*	*0.0002*	
C	*12.25*	*1*	*12.25*	*18.59*	*0.0012*	
AB	*2.25*	*1*	*2.25*	*3.41*	*0.0917*	
Residual	7.25	11	0.66			
Lack of Fit	*2.25*	*3*	*0.75*	*1.20*	*0.3700*	*not significant*
Pure Error	*5.00*	*8*	*0.63*			
Cor Total	78.00	15				

The Model F-value of 26.84 implies the model is significant. There is only
a 0.01% chance that a "Model F-Value" this large could occur due to noise.

Values of "Prob > F" less than 0.0500 indicate model terms are significant.
In this case A, B, C are significant model terms.

Std. Dev.	0.81	R-Squared	0.9071
Mean	1.00	Adj R-Squared	0.8733
C.V.	81.18	Pred R-Squared	0.8033
PRESS	15.34	Adeq Precision	15.424

(b) Analyze the residuals from this experiment. Are there any indications of model inadequacy?

The residual plots below do not identify any violations to the assumptions.

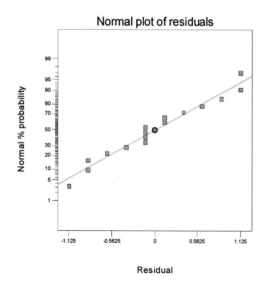

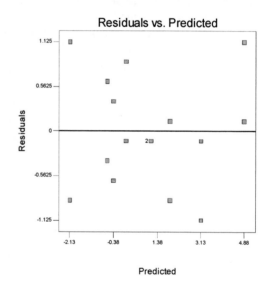

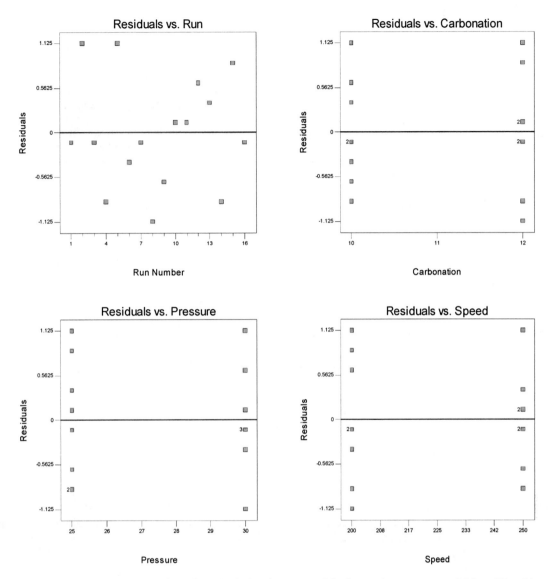

(c) Obtain a model for predicting fill height deviation in terms of the important process variables. Use this model to construct contour plots to assist in interpreting the results of the experiment.

The model in both coded and actual factors are shown below.

Design-Expert Output

Final Equation in Terms of Coded Factors:

        Fill Deviation   =
                +1.00
                +1.50   * A
                +1.13   * B
                +0.88   * C
                +0.38   * A * B

Final Equation in Terms of Actual Factors:

Fill Deviation =
+9.62500
-2.62500 * Carbonation
-1.20000 * Pressure
+0.035000 * Speed
+0.15000 * Carbonation * Pressure

The following contour plots identify the fill deviation with respect to carbonation and pressure. The plot on the left sets the speed at 200 b/m while the plot on the right sets the speed at 250 b/m. Assuming a faster bottle speed is better, settings on pressure and carbonation that produce a fill deviation near zero can be found in the lower left hand corner of the contour plot on the right.

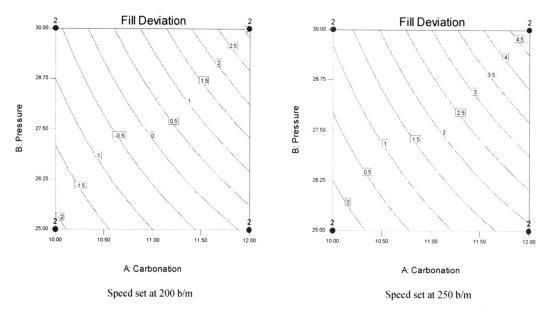

Speed set at 200 b/m   Speed set at 250 b/m

(d) In part (a), you probably noticed that there was an interaction term that was borderline significant. If you did not include the interaction term in your model, include it now and repeat the analysis. What difference did this make? If you elected to include the interaction term in part (a), remove it and repeat the analysis. What difference does this make?

The following analysis of variance, residual plots, and contour plots represent the model without the interaction. As in the original analysis, the residual plots do not identify any concerns with the assumptions. The contour plots did not change significantly either. The interaction effect is small relative to the main effects.

Design-Expert Output

Response: Fill Deviation
  ANOVA for Selected Factorial Model
Analysis of variance table [Partial sum of squares]

Source	Sum of Squares	DF	Mean Square	F Value	Prob > F	
Model	68.50	3	22.83	28.84	< 0.0001	significant
A	36.00	1	36.00	45.47	< 0.0001	
B	20.25	1	20.25	25.58	0.0003	
C	12.25	1	12.25	15.47	0.0020	
Residual	9.50	12	0.79			
Lack of Fit	4.50	4	1.13	1.80	0.2221	not significant

Pure Error	5.00	8	0.63
Cor Total	78.00	15	

The Model F-value of 28.84 implies the model is significant. There is only
a 0.01% chance that a "Model F-Value" this large could occur due to noise.

Values of "Prob > F" less than 0.0500 indicate model terms are significant.
In this case A, B, C are significant model terms.

Std. Dev.	0.89	R-Squared	0.8782
Mean	1.00	Adj R-Squared	0.8478
C.V.	88.98	Pred R-Squared	0.7835
PRESS	16.89	Adeq Precision	15.735

Final Equation in Terms of Coded Factors:

Fill Deviation =
   +1.00
   +1.50   * A
   +1.13   * B
   +0.88   * C

Final Equation in Terms of Actual Factors:

Fill Deviation =
   -35.75000
   +1.50000   * Carbonation
   +0.45000   * Pressure
   +0.035000   * Speed

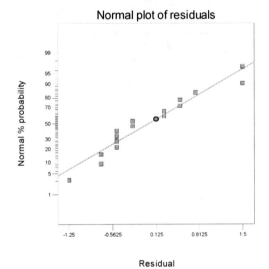

Normal plot of residuals

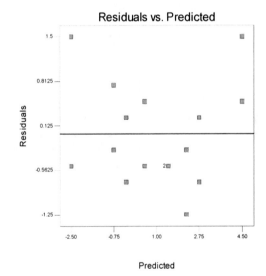

Residuals vs. Predicted

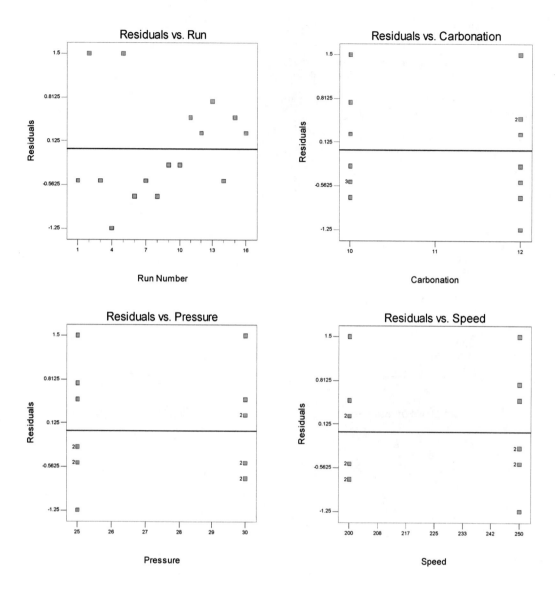

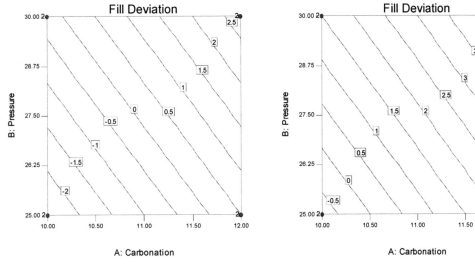

| Speed set at 200 b/m | Speed set at 250 b/m |

6.21. I am always interested in improving my golf scores. Since a typical golfer uses the putter for about 35-45% of his or her strokes, it seems reasonable that improving one's putting is a logical and perhaps simple way to improve a golf score ("The man who can putt is a match for any man." – Willie Parks, 1864-1925, two-time winner of the British Open). An experiment was conducted to study the effects of four factors on putting accuracy. The design factors are length of putt, type of putter, breaking putt vs. straight putt, and level versus downhill putt. The response variable is distance from the ball to the center of the cup after the ball comes to rest. One golfer performs the experiment, a 2^4 factorial design with seven replicates was used, and all putts were made in random order. The results are sjpwm om Table P6.4.

Table P6.4

Design Factors				Distance from cup (replicates)						
Length of putt (ft)	Type of putter	Break of putt	Slope of putt	1	2	3	4	5	6	7
10	Mallet	Straight	Level	10.0	18.0	14.0	12.5	19.0	16.0	18.5
30	Mallet	Straight	Level	0.0	16.5	4.5	17.5	20.5	17.5	33.0
10	Cavity-back	Straight	Level	4.0	6.0	1.0	14.5	12.0	14.0	5.0
30	Cavity-back	Straight	Level	0.0	10.0	34.0	11.0	25.5	21.5	0.0
10	Mallet	Breaking	Level	0.0	0.0	18.5	19.5	16.0	15.0	11.0
30	Mallet	Breaking	Level	5.0	20.5	18.0	20.0	29.5	19.0	10.0
10	Cavity-back	Breaking	Level	6.5	18.5	7.5	6.0	0.0	10.0	0.0
30	Cavity-back	Breaking	Level	16.5	4.5	0.0	23.5	8.0	8.0	8.0
10	Mallet	Straight	Downhill	4.5	18.0	14.5	10.0	0.0	17.5	6.0
30	Mallet	Straight	Downhill	19.5	18.0	16.0	5.5	10.0	7.0	36.0
10	Cavity-back	Straight	Downhill	15.0	16.0	8.5	0.0	0.5	9.0	3.0
30	Cavity-back	Straight	Downhill	41.5	39.0	6.5	3.5	7.0	8.5	36.0
10	Mallet	Breaking	Downhill	8.0	4.5	6.5	10.0	13.0	41.0	14.0
30	Mallet	Breaking	Downhill	21.5	10.5	6.5	0.0	15.5	24.0	16.0
10	Cavity-back	Breaking	Downhill	0.0	0.0	0.0	4.5	1.0	4.0	6.5
30	Cavity-back	Breaking	Downhill	18.0	5.0	7.0	10.0	32.5	18.5	8.0

(a) Analyze the data from this experiment. Which factors significantly affect putting performance?

The half normal probability plot of effects identifies only factors A and B, length of putt and type of putter, as having a potentially significant effect on putting performance. The analysis of variance with only these significant factors is presented as well and confirms significance.

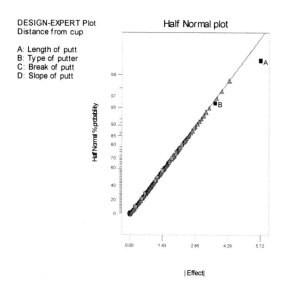

DESIGN-EXPERT Plot
Distance from cup

A: Length of putt
B: Type of putter
C: Break of putt
D: Slope of putt

Half Normal plot

Design-Expert Output with Only Factors A and B

Response: Distance from cup
ANOVA for Selected Factorial Model
Analysis of variance table [Terms added sequentially (first to last)]

Source	Sum of Squares	DF	Mean Square	F Value	Prob > F	
Model	1305.29	2	652.65	7.69	0.0008	significant
A	917.15	1	917.15	10.81	0.0014	
B	388.15	1	388.15	4.57	0.0347	
Residual	9248.94	109	84.85			
Lack of Fit	933.15	13	71.78	0.83	0.6290	not significant
Pure Error	8315.79	96	86.62			
Cor Total	10554.23	111				

The Model F-value of 7.69 implies the model is significant. There is only a 0.08% chance that a "Model F-Value" this large could occur due to noise.

Values of "Prob > F" less than 0.0500 indicate model terms are significant. In this case A, B, are significant model terms.

Std. Dev.	9.21	R-Squared	0.1237
Mean	12.30	Adj R-Squared	0.1076
C.V.	74.90	Pred R-Squared	0.0748
PRESS	9765.06	Adeq Precision	6.266

(b) Analyze the residuals from this experiment. Are there any indications of model inadequacy?

The residual plots for the model containing only the significant factors A and B are shown below. The normality assumption appears to be violated. Also, as a golfer might expect, there is a slight inequality of variance with regards to the length of putt. A square root transformation is applied which corrects the violations. The analysis of variance and corrected residual plots are also presented. Finally, an effects plot identifies a 10 foot putt and the cavity-back putter reduce the mean distance from the cup.

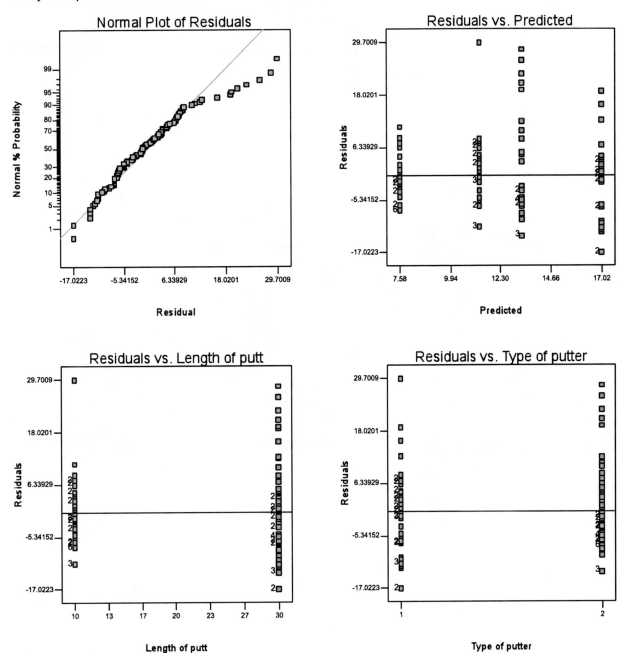

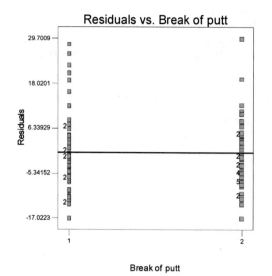

Residuals vs. Break of putt

Break of putt

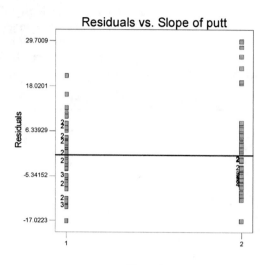

Residuals vs. Slope of putt

Slope of putt

Design-Expert Output with Only Factors A and B and a Square Root Transformation

Response: Distance from cup   **Transform: Square root**   **Constant: 0**
   ANOVA for Selected Factorial Model
Analysis of variance table [Terms added sequentially (first to last)]

Source	Sum of Squares	DF	Mean Square	F Value	Prob > F	
Model	37.26	2	18.63	7.85	0.0007	significant
A	21.61	1	21.61	9.11	0.0032	
B	15.64	1	15.64	6.59	0.0116	
Residual	258.63	109	2.37			
Lack of Fit	30.19	13	2.32	0.98	0.4807	not significant
Pure Error	228.45	96	2.38			
Cor Total	295.89	111				

The Model F-value of 7.85 implies the model is significant. There is only
a 0.07% chance that a "Model F-Value" this large could occur due to noise.

Values of "Prob > F" less than 0.0500 indicate model terms are significant.
In this case A, B, are significant model terms.

Std. Dev.	1.54	R-Squared	0.1259	
Mean	3.11	Adj R-Squared	0.1099	
C.V.	49.57	Pred R-Squared	0.0771	
PRESS	273.06	Adeq Precision	6.450	

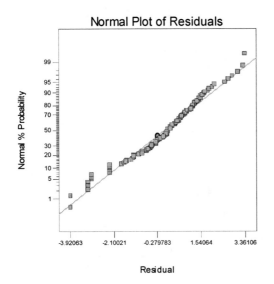

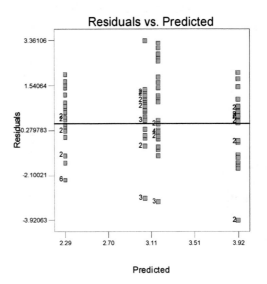

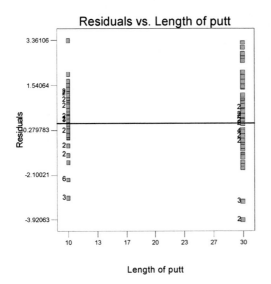

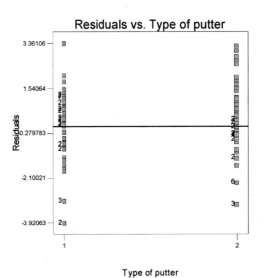

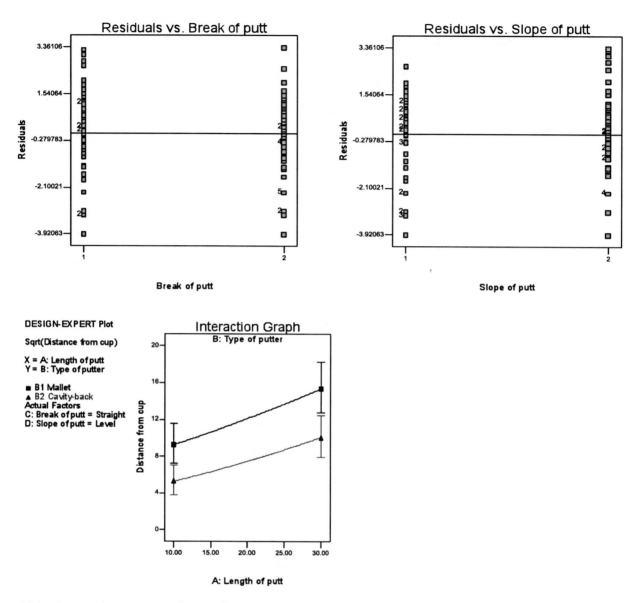

6.26. An experiment was run in a semiconductor fabrication plant in an effort to increase yield. Five factors, each at two levels, were studied. The factors (and levels) were A = aperture setting (small, large), B = exposure time (20% below nominal, 20% above nominal), C = development time (30 s, 45 s), D = mask dimension (small, large), and E = etch time (14.5 min, 15.5 min). The unreplicated 2^5 design shown below was run.

(1) =	7	d =	8	e =	8	de =	6
a =	9	ad =	10	ae =	12	ade =	10
b =	34	bd =	32	be =	35	bde =	30
ab =	55	abd =	50	abe =	52	$abde$ =	53
c =	16	cd =	18	ce =	15	cde =	15
ac =	20	acd =	21	ace =	22	$acde$ =	20
bc =	40	bcd =	44	bce =	45	$bcde$ =	41
abc =	60	$abcd$ =	61	$abce$ =	65	$abcde$ =	63

(a) Construct a normal probability plot of the effect estimates. Which effects appear to be large?

From the normal probability plot of effects shown below, effects *A*, *B*, *C*, and the *AB* interaction appear to be large.

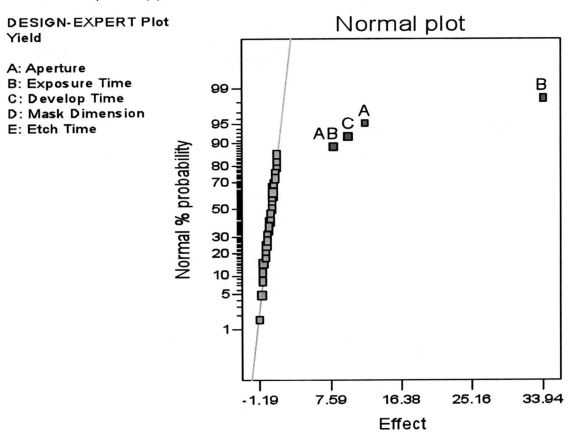

(b) Conduct an analysis of variance to confirm your findings for part (a).

Design-Expert Output

Response: Yield
    ANOVA for Selected Factorial Model
Analysis of variance table [Partial sum of squares]

Source	Sum of Squares	DF	Mean Square	F Value	Prob > F	
Model	11585.13	4	2896.28	991.83	< 0.0001	significant
A	1116.28	1	1116.28	382.27	< 0.0001	
B	9214.03	1	9214.03	3155.34	< 0.0001	
C	750.78	1	750.78	257.10	< 0.0001	
AB	504.03	1	504.03	172.61	< 0.0001	
Residual	78.84	27	2.92			
Cor Total	11663.97	31				

The Model F-value of 991.83 implies the model is significant. There is only
a 0.01% chance that a "Model F-Value" this large could occur due to noise.

Values of "Prob > F" less than 0.0500 indicate model terms are significant.
In this case A, B, C, AB are significant model terms.

(c) Write down the regression model relating yield to the significant process variables.

Design-Expert Output

Final Equation in Terms of Actual Factors:

Aperture   small

    Yield   =
    +0.40625
    +0.65000   * Exposure Time
    +0.64583   * Develop Time

Aperture   large

    Yield   =
    +12.21875
    +1.04688   * Exposure Time
    +0.64583   * Develop Time

(d) Plot the residuals on normal probability paper. Is the plot satisfactory?

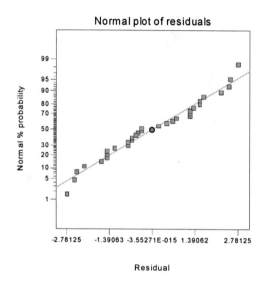

There is nothing unusual about this plot.

(e) Plot the residuals versus the predicted yields and versus each of the five factors. Comment on the plots.

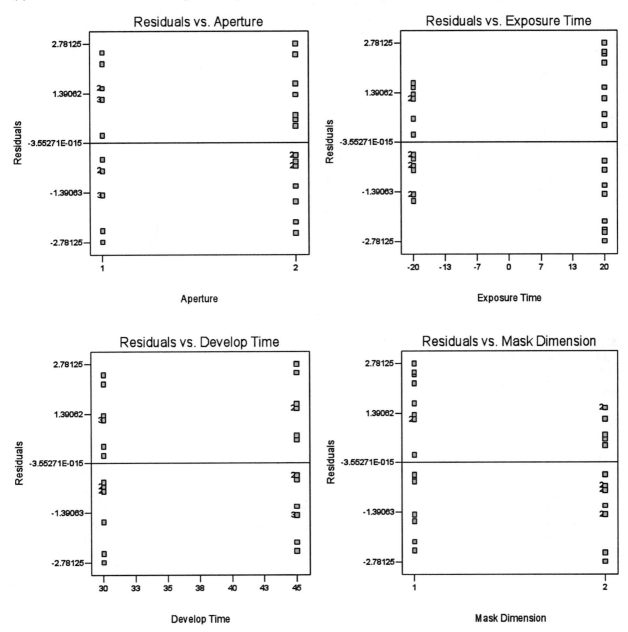

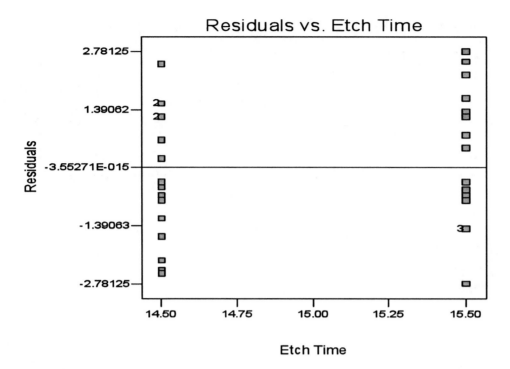

The plot of residual versus exposure time shows some very slight inequality of variance. There is no strong evidence of a potential problem.

(f) Interpret any significant interactions.

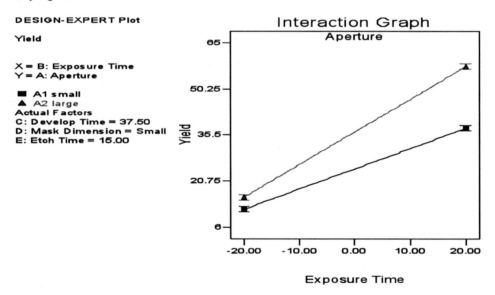

Factor *A* does not have as large an effect when *B* is at its low level as it does when *B* is at its high level.

(g) What are your recommendations regarding process operating conditions?

To achieve the highest yield, run *B* at the high level, *A* at the high level, and *C* at the high level.

(h) Project the 2^5 design in this problem into a 2^k design in the important factors. Sketch the design and show the average and range of yields at each run. Does this sketch aid in interpreting the results of this experiment?

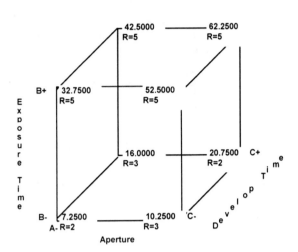

DESIGN-EASE Analysis
Actual Yield

This cube plot aids in interpretation. The strong AB interaction and the large positive effect of C are clearly evident.

6.28. In a process development study on yield, four factors were studied, each at two levels: time (A), concentration (B), pressure (C), and temperature (D). A single replicate of a 2^4 design was run, and the resulting data are shown in Table P6.7.

Table P6.7

Run Number	Actual Run Order	A	B	C	D	Yield (lbs)	Factor	Levels Low (-)	High (+)
1	5	-	-	-	-	12	A (h)	2.5	3.0
2	9	+	-	-	-	18	B (%)	14	18
3	8	-	+	-	-	13	C (psi)	60	80
4	13	+	+	-	-	16	D (°C)	225	250
5	3	-	-	+	-	17			
6	7	+	-	+	-	15			
7	14	-	+	+	-	20			
8	1	+	+	+	-	15			
9	6	-	-	-	+	10			
10	11	+	-	-	+	25			
11	2	-	+	-	+	13			
12	15	+	+	-	+	24			
13	4	-	-	+	+	19			
14	16	+	-	+	+	21			
15	10	-	+	+	+	17			
16	12	+	+	+	+	23			

(a) Construct a normal probability plot of the effect estimates. Which factors appear to have large effects?

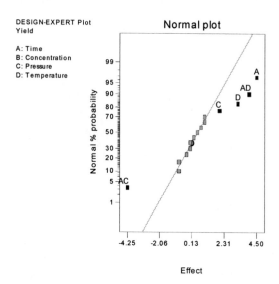

A, C, D and the AC and AD interactions appear to have large effects.

(b) Conduct an analysis of variance using the normal probability plot in part (a) for guidance in forming an error term. What are your conclusions?

Design-Expert Output

Response: Yield
ANOVA for Selected Factorial Model
Analysis of variance table [Partial sum of squares]

Source	Sum of Squares	DF	Mean Square	F Value	Prob > F	
Model	275.50	5	55.10	33.91	< 0.0001	significant
A	81.00	1	81.00	49.85	< 0.0001	
C	16.00	1	16.00	9.85	0.0105	
D	42.25	1	42.25	26.00	0.0005	
AC	72.25	1	72.25	44.46	< 0.0001	
AD	64.00	1	64.00	39.38	< 0.0001	
Residual	16.25	10	1.62			
Cor Total	291.75	15				

The Model F-value of 33.91 implies the model is significant. There is only a 0.01% chance that a "Model F-Value" this large could occur due to noise.

Values of "Prob > F" less than 0.0500 indicate model terms are significant. In this case A, C, D, AC, AD are significant model terms.

(c) Write down a regression model relating yield to the important process variables.

Design-Expert Output

Final Equation in Terms of Coded Factors:

Yield =
+17.38
+2.25 *A
+1.00 *C
+1.63 *D
-2.13 *A*C
+2.00 *A*D

Final Equation in Terms of Actual Factors:

Yield =
+209.12500
-83.50000 * Time
+2.43750 * Pressure
-1.63000 * Temperature
-0.85000 * Time * Pressure
+0.64000 * Time * Temperature

(d) Analyze the residuals from this experiment. Does your analysis indicate any potential problems?

From the following residual plots, there is no indication of problems with assumptions or model adequacy.

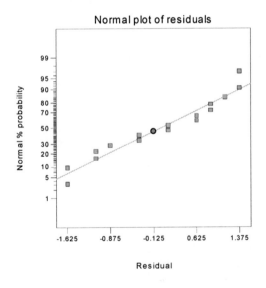

Normal plot of residuals

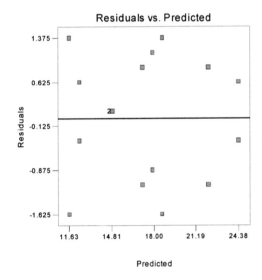

Residuals vs. Predicted

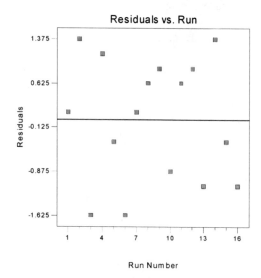

Residuals vs. Run

(e) Can this design be collapsed into a 2^3 design with two replicates? If so, sketch the design with the average and range of yield shown at each point in the cube. Interpret the results.

With the removal of factor B, the design is collapsed into a 2^3 design with two replicates as shown in the following diagram.

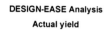

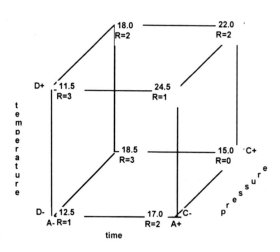

6.29. Continuation of Problem 6.28. Use the regression model in part (c) of Problem 6-28 to generate a response surface contour plot of yield. Discuss the practical value of this response surface plot.

The response surface contour plot shows the adjustments in the process variables that lead to an increasing or decreasing response. It also displays the curvature of the response in the design region, possibly indicating where robust operating conditions can be found. Two response surface contour plots for this process are shown below. These were formed from the model written in terms of the original design variables.

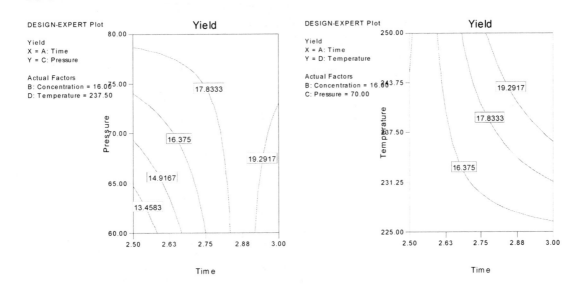

6.31. An experiment was conducted on a chemical process that produces a polymer. The four factors studied were temperature (A), catalyst concentration (B), time (C), and pressure (D). Two responses, molecular weight and viscosity, were observed. The design matrix and response data in Table P6.8.

Table P6.8

Run Number	Actual Run Order	A	B	C	D	Molecular Weight	Viscosity		Factor Low (-)	Levels High (+)
1	18	-	-	-	-	2400	1400	A (°C)	100	120
2	9	+	-	-	-	2410	1500	B (%)	4	8
3	13	-	+	-	-	2315	1520	C (min)	20	30
4	8	+	+	-	-	2510	1630	D (psi)	60	75
5	3	-	-	+	-	2615	1380			
6	11	+	-	+	-	2625	1525			
7	14	-	+	+	-	2400	1500			
8	17	+	+	+	-	2750	1620			
9	6	-	-	-	+	2400	1400			
10	7	+	-	-	+	2390	1525			
11	2	-	+	-	+	2300	1500			
12	10	+	+	-	+	2520	1500			
13	4	-	-	+	+	2625	1420			
14	19	+	-	+	+	2630	1490			
15	15	-	+	+	+	2500	1500			

16	20	+	+	+	+	2710	1600
17	1	0	0	0	0	2515	1500
18	5	0	0	0	0	2500	1460
19	16	0	0	0	0	2400	1525
20	12	0	0	0	0	2475	1500

(a) Consider only the molecular weight response. Plot the effect estimates on a normal probability scale. What effects appear important?

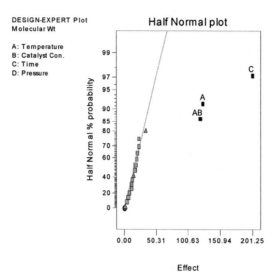

The effects for factors A, C, and the AB interaction appear to be significant.

(b) Use an analysis of variance to confirm the results from part (a). Is there an indication of curvature?

A,C and the AB interaction are significant. While the main effect of B is not significant, it could be included to preserve hierarchy in the model. There is no indication of quadratic curvature.

Design-Expert Output

Response: Molecular Wt
ANOVA for Selected Factorial Model
Analysis of variance table [Partial sum of squares]

Source	Sum of Squares	DF	Mean Square	F Value	Prob > F	
Model	2.809E+005	3	93620.83	73.00	< 0.0001	significant
A	61256.25	1	61256.25	47.76	< 0.0001	
C	1.620E+005	1	1.620E+005	126.32	< 0.0001	
AB	57600.00	1	57600.00	44.91	< 0.0001	
Curvature	3645.00	1	3645.00	2.84	0.1125	not significant
Residual	19237.50	15	1282.50			
Lack of Fit	11412.50	12	951.04	0.36	0.9106	not significant
Pure Error	7825.00	3	2608.33			
Cor Total	3.037E+005	19				

The Model F-value of 73.00 implies the model is significant. There is only a 0.01% chance that a "Model F-Value" this large could occur due to noise.

Values of "Prob > F" less than 0.0500 indicate model terms are significant.
In this case A, C, AB are significant model terms.

(c) Write down a regression model to predict molecular weight as a function of the important variables.

Design-Expert Output

Final Equation in Terms of Coded Factors:
Molecular Wt   =
+2506.25
+61.87   * A
+100.63   * C
+60.00   * A * B

(d) Analyze the residuals and comment on model adequacy.

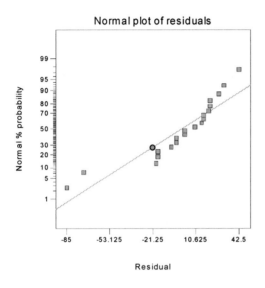

Normal plot of residuals

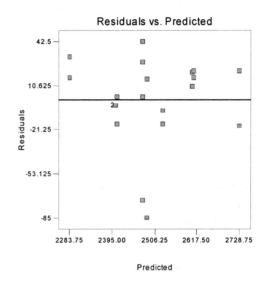

Residuals vs. Predicted

There are two residuals that appear to be large and should be investigated.

(e) Repeat parts (a) - (d) using the viscosity response.

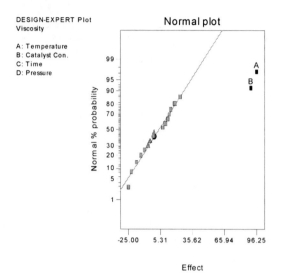

Design-Expert Output

Response: Viscosity
    ANOVA for Selected Factorial Model
Analysis of variance table [Partial sum of squares]

Source	Sum of Squares	DF	Mean Square	F Value	Prob > F	
Model	70362.50	2	35181.25	35.97	< 0.0001	significant
A	37056.25	1	37056.25	37.88	< 0.0001	
B	33306.25	1	33306.25	34.05	< 0.0001	
Curvature	61.25	1	61.25	0.063	0.8056	not significant
Residual	15650.00	16	978.13			
Lack of Fit	13481.25	13	1037.02	1.43	0.4298	not significant
Pure Error	2168.75	3	722.92			
Cor Total	86073.75	19				

The Model F-value of 35.97 implies the model is significant. There is only
a 0.01% chance that a "Model F-Value" this large could occur due to noise.

Values of "Prob > F" less than 0.0500 indicate model terms are significant.
In this case A, B are significant model terms.

Final Equation in Terms of Coded Factors:

    Viscosity =
    +1500.62
      +48.13 * A
      +45.63 * B

Normal plot of residuals

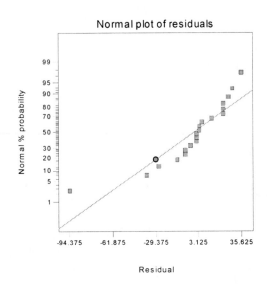

Residuals vs. Predicted

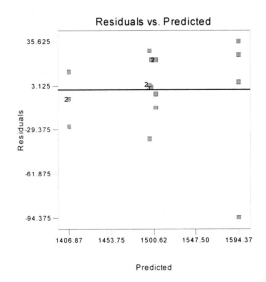

There is one large residual that should be investigated.

6.32. Continuation of Problem 6.31. Use the regression models for molecular weight and viscosity to answer the following questions.

(a) Construct a response surface contour plot for molecular weight. In what direction would you adjust the process variables to increase molecular weight?

Increase temperature, catalyst and time.

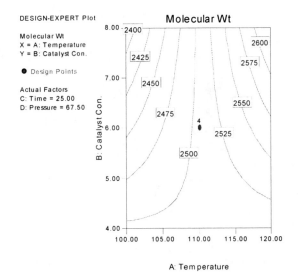

(b) Construct a response surface contour plot for viscosity. In what direction would you adjust the process variables to decrease viscosity?

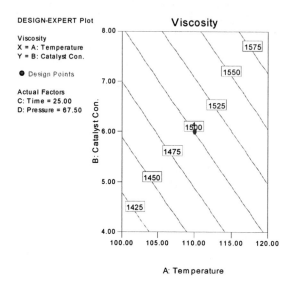

Decreasing temperature and catalyst would decrease viscosity.

(c) What operating conditions would you recommend if it was necessary to produce a product with a molecular weight between 2400 and 2500, and the lowest possible viscosity?

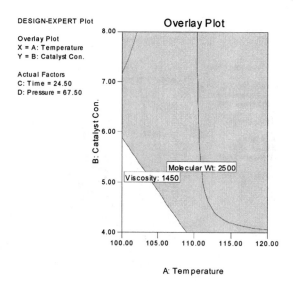

Set the temperature between 100 and 105, the catalyst between 4 and 5%, and the time at 24.5 minutes. The pressure was not significant and can be set at conditions that may improve other results of the process, such as cost.

6.36. Resistivity on a silicon wafer is influenced by several factors. The results of a 2^4 factorial experiment performed during a critical process step is shown in Table P6.10.

Table P6.10

Run	A	B	C	D	Resistivity
1	-	-	-	-	1.92
2	+	-	-	-	11.28
3	-	+	-	-	1.09
4	+	+	-	-	5.75
5	-	-	+	-	2.13
6	+	-	+	-	9.53
7	-	+	+	-	1.03
8	+	+	+	-	5.35
9	-	-	-	+	1.60
10	+	-	-	+	11.73
11	-	+	-	+	1.16
12	+	+	-	+	4.68
13	-	-	+	+	2.16
14	+	-	+	+	9.11
15	-	+	+	+	1.07
16	+	+	+	+	5.30

(a) Estimate the factor effects. Plot the effect estimates on a normal probability plot and select a tentative model.

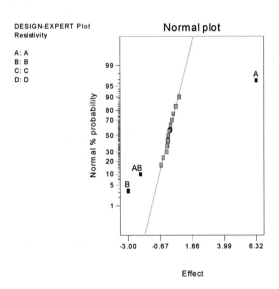

(b) Fit the model identified in part (a) and analyze the residuals. Is there any indication of model inadequacy?

The normal probability plot of residuals is not satisfactory. The plots of residual versus predicted, residual versus factor *A*, and residual versus factor *B* are funnel shaped, indicating non-constant variance.

Design-Expert Output

Response: Resistivity
ANOVA for Selected Factorial Model
Analysis of variance table [Partial sum of squares]

Source	Sum of Squares	DF	Mean Square	F Value	Prob > F	
Model	214.22	3	71.41	148.81	< 0.0001	significant
A	159.83	1	159.83	333.09	< 0.0001	
B	36.09	1	36.09	75.21	< 0.0001	
AB	18.30	1	18.30	38.13	< 0.0001	
Residual	5.76	12	0.48			
Cor Total	219.98	15				

The Model F-value of 148.81 implies the model is significant. There is only
a 0.01% chance that a "Model F-Value" this large could occur due to noise.

Values of "Prob > F" less than 0.0500 indicate model terms are significant.
In this case A, B, AB are significant model terms.

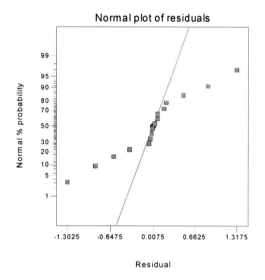

Normal plot of residuals

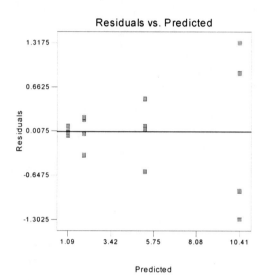

Residuals vs. Predicted

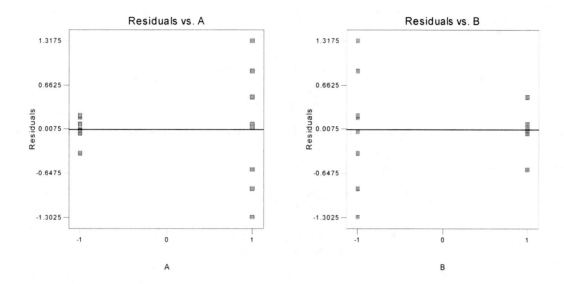

(c) Repeat the analysis from parts (a) and (b) using ln(*y*) as the response variable. Is there any indication that the transformation has been useful?

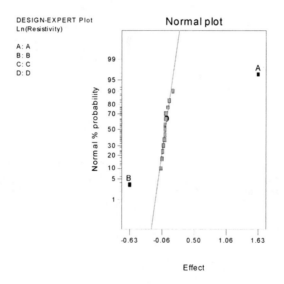

Design-Expert Output

Response: Resistivity	Transform: Natural log		Constant:	0.000		

ANOVA for Selected Factorial Model

Analysis of variance table [Partial sum of squares]

Source	Sum of Squares	DF	Mean Square	F Value	Prob > F	
Model	12.15	2	6.08	553.44	< 0.0001	significant
A	10.57	1	10.57	962.95	< 0.0001	
B	1.58	1	1.58	143.94	< 0.0001	
Residual	0.14	13	0.011			
Cor Total	12.30	15				

The Model F-value of 553.44 implies the model is significant. There is only
a 0.01% chance that a "Model F-Value" this large could occur due to noise.

Values of "Prob > F" less than 0.0500 indicate model terms are significant.
In this case A, B are significant model terms.

The transformed data no longer indicates that the AB interaction is significant. A simpler model has resulted from the log transformation. The residual plots shown below are much improved.

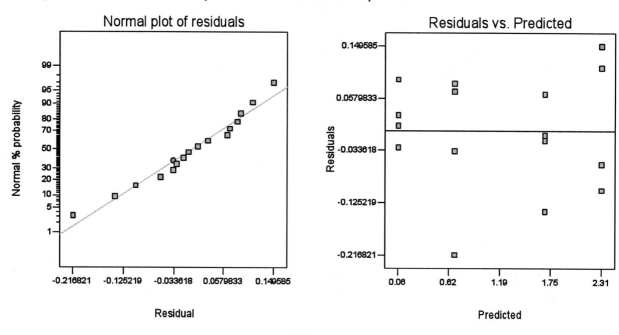

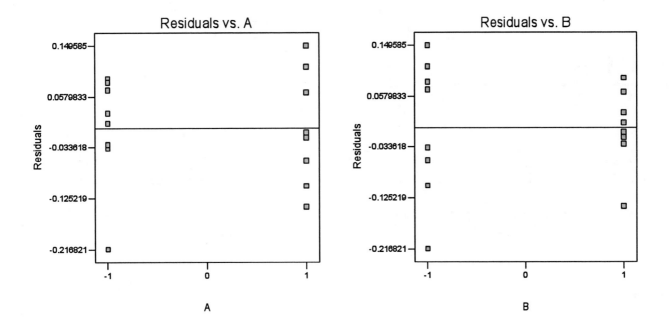

(d) Fit a model in terms of the coded variables that can be used to predict the resistivity.

Design-Expert Output

Final Equation in Terms of Coded Factors:

Ln(Resistivity)   =
    +1.19
    +0.81   * A
    -0.31   * B

CHAPTER 7

Blocking and Confounding in the 2^k Factorial Design

LEARNING OBJECTIVES

After completing this chapter, you will be able to:

1. Know how to conduct a replicated 2^k design in blocks.

2. Know how to conduct an unreplicated 2^k design in blocks by confounding certain interactions with the blocks.

3. Perform the statistical analysis for a blocked 2^k factorial design.

KEY CONCEPTS AND IDEAS

1. Blocking

2. Confounding

3. Defining contrast

4. Partial confounding

5. Generalized interaction

Exercises

7.1. Consider the experiment described in Problem 6.1. Analyze this experiment assuming that each replicate represents a block of a single production shift.

Source of Variation	Sum of Squares	Degrees of Freedom	Mean Square	F_0
Cutting Speed (A)	0.67	1	0.67	<1
Tool Geometry (B)	770.67	1	770.67	22.38*
Cutting Angle (C)	280.17	1	280.17	8.14*
AB	16.67	1	16.67	<1
AC	468.17	1	468.17	13.60*
BC	48.17	1	48.17	1.40
ABC	28.17	1	28.17	<1
Blocks	0.58	2	0.29	
Error	482.08	14	34.43	
Total	2095.33	23		

Design-Expert Output

Response: Life in hours
  ANOVA for Selected Factorial Model
Analysis of variance table [Partial sum of squares]

Source	Sum of Squares	DF	Mean Square	F Value	Prob > F	
Block	0.58	2	0.29			
Model	1519.67	4	379.92	11.23	0.0001	significant
A	0.67	1	0.67	0.020	0.8900	
B	770.67	1	770.67	22.78	0.0002	
C	280.17	1	280.17	8.28	0.0104	
AC	468.17	1	468.17	13.84	0.0017	
Residual	575.08	17	33.83			
Cor Total	2095.33	23				

The Model F-value of 11.23 implies the model is significant. There is only a 0.01% chance that a "Model F-Value" this large could occur due to noise.

Values of "Prob > F" less than 0.0500 indicate model terms are significant.
In this case B, C, AC are significant model terms.

These results agree with the results from Problem 6.1. Tool geometry, cutting angle and the interaction between cutting speed and cutting angle are significant at the 5% level. The Design-Expert program also includes factor A, cutting speed, in the model to preserve hierarchy.

7.4. Consider the data from the first replicate of Problem 6.1. Suppose that these observations could not all be run using the same bar stock. Set up a design to run these observations in two blocks of four observations each with *ABC* confounded. Analyze the data.

Block 1	Block 2
(1)	a
ab	b
ac	c
bc	abc

From the normal probability plot of effects, *B*, *C*, and the *AC* interaction are significant. Factor *A* was included in the analysis of variance to preserve hierarchy.

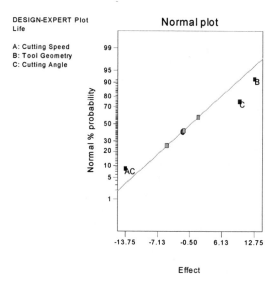

Design-Expert Output

Response: Life in hours
ANOVA for Selected Factorial Model
Analysis of variance table [Partial sum of squares]

Source	Sum of Squares	DF	Mean Square	F Value	Prob > F	
Block	91.13	1	91.13			
Model	896.50	4	224.13	7.32	0.1238	not significant
A	3.13	1	3.13	0.10	0.7797	
B	325.12	1	325.12	10.62	0.0827	
C	190.12	1	190.12	6.21	0.1303	
AC	378.13	1	378.13	12.35	0.0723	
Residual	61.25	2	30.62			
Cor Total	1048.88	7				

The "Model F-value" of 7.32 implies the model is not significant relative to the noise. There is a
12.38 % chance that a "Model F-value" this large could occur due to noise.

Values of "Prob > F" less than 0.0500 indicate model terms are significant.
In this case there are no significant model terms.

This design identifies the same significant factors as Problem 6.1.

7.7. Using the data from the 2^5 design in Problem 6.24, construct and analyze a design in two blocks with *ABCDE* confounded with blocks.

Block 1	Block 1	Block 2	Block 2
(1)	ae	a	e
ab	be	b	abe
ac	ce	c	ace
bc	abce	abc	bce
ad	de	d	ade
bd	abde	abd	bde
cd	acde	acd	cde
abcd	bcde	bcd	abcde

The normal probability plot of effects identifies factors *A*, *B*, *C*, and the *AB* interaction as being significant. This is confirmed with the analysis of variance.

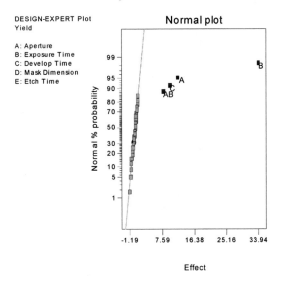

Design-Expert Output

Response: Yield
    ANOVA for Selected Factorial Model
    Analysis of variance table [Partial sum of squares]

Source	Sum of Squares	DF	Mean Square	F Value	Prob > F	
Block	0.28	1	0.28			
Model	11585.13	4	2896.28	958.51	< 0.0001	significant
A	1116.28	1	1116.28	369.43	< 0.0001	
B	9214.03	1	9214.03	3049.35	< 0.0001	
C	750.78	1	750.78	248.47	< 0.0001	
AB	504.03	1	504.03	166.81	< 0.0001	
Residual	78.56	26	3.02			

Cor Total	11663.97	31

The Model F-value of 958.51 implies the model is significant. There is only a 0.01% chance that a "Model F-Value" this large could occur due to noise.

Values of "Prob > F" less than 0.0500 indicate model terms are significant. In this case A, B, C, AB are significant model terms.

7.8 Repeat Problem 7.7 assuming that four blocks are necessary. Suggest a reasonable confounding scheme.

Use *ABC* and *CDE*, and consequently *ABDE*. The four blocks follow.

Block 1	Block 2	Block 3	Block 4
(1)	*a*	*ac*	*c*
ab	*b*	*bc*	*abc*
acd	*cd*	*d*	*ad*
bcd	*abcd*	*abd*	*bd*
ace	*ce*	*e*	*ae*
bce	*abce*	*abe*	*be*
de	*ade*	*acde*	*cde*
abde	*bde*	*bcde*	*abcde*

The half normal probability plot of effects identifies the same significant effects as in Problem 7.7.

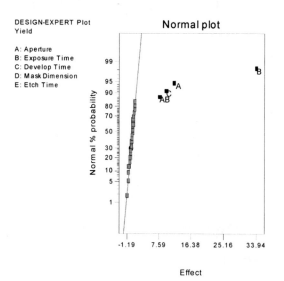

Design-Expert Output

Response: Yield
      ANOVA for Selected Factorial Model
    Analysis of variance table [Partial sum of squares]

Source	Sum of Squares	DF	Mean Square	F Value	Prob > F	
Block	13.84	3	4.61			
Model	11585.13	4	2896.28	1069.40	< 0.0001	significant
A	1116.28	1	1116.28	412.17	< 0.0001	
B	9214.03	1	9214.03	3402.10	< 0.0001	
C	750.78	1	750.78	277.21	< 0.0001	
AB	504.03	1	504.03	186.10	< 0.0001	
Residual	65.00	24	2.71			
Cor Total	11663.97	31				

The Model F-value of 1069.40 implies the model is significant. There is only
a 0.01% chance that a "Model F-Value" this large could occur due to noise.

Values of "Prob > F" less than 0.0500 indicate model terms are significant.
In this case A, B, C, AB are significant model terms.

7.9. Consider the data from the 2^5 design in Problem 6.26. Suppose that it was necessary to run this design in four blocks with *ACDE* and *BCD* (and consequently *ABE*) confounded. Analyze the data from this design.

Block 1	Block 2	Block 3	Block 4
(1)	a	b	c
ae	e	abe	ace
cd	acd	bcd	d
abc	bc	ac	ab
acde	cde	abcde	ade
bce	abce	ce	be
abd	bd	ad	abcd
bde	abde	de	bcde

Even with four blocks, the same effects are identified as significant in the following normal probability plot and analysis of variance.

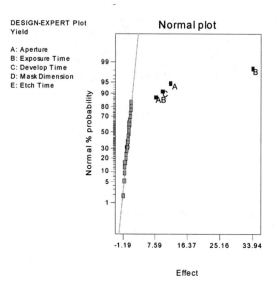

DESIGN-EXPERT Plot
Yield

A: Aperture
B: Exposure Time
C: Develop Time
D: Mask Dimension
E: Etch Time

Design-Expert Output

Response: Yield
   ANOVA for Selected Factorial Model
Analysis of variance table [Partial sum of squares]

Source	Sum of Squares	DF	Mean Square	F Value	Prob > F	
Block	2.59	3	0.86			
Model	11585.13	4	2896.28	911.62	< 0.0001	significant
A	1116.28	1	1116.28	351.35	< 0.0001	
B	9214.03	1	9214.03	2900.15	< 0.0001	
C	750.78	1	750.78	236.31	< 0.0001	
AB	504.03	1	504.03	158.65	< 0.0001	
Residual	76.25	24	3.18			
Cor Total	11663.97	31				

The Model F-value of 911.62 implies the model is significant. There is only
a 0.01% chance that a "Model F-Value" this large could occur due to noise.

Values of "Prob > F" less than 0.0500 indicate model terms are significant.
In this case A, B, C, AB are significant model terms.

7.10. Consider the fill height deviation experiment in Problem 6.20. Suppose that each replicate was run on a separate day. Analyze the data assuming that the days are blocks.

The shown below analysis is very similar to the original analysis in chapter 6. The same effects are significant.

Design-Expert Output

Response: Fill Deviation
ANOVA for Selected Factorial Model
Analysis of variance table [Partial sum of squares]

Source	Sum of Squares	DF	Mean Square	F Value	Prob > F	
Block	1.00	1	1.00			
Model	70.75	4	17.69	28.30	< 0.0001	significant
A	36.00	1	36.00	57.60	< 0.0001	
B	20.25	1	20.25	32.40	0.0002	
C	12.25	1	12.25	19.60	0.0013	
AB	2.25	1	2.25	3.60	0.0870	
Residual	6.25	10	0.62			
Cor Total	78.00	15				

The Model F-value of 28.30 implies the model is significant. There is only
a 0.01% chance that a "Model F-Value" this large could occur due to noise.

Values of "Prob > F" less than 0.0500 indicate model terms are significant.
In this case A, B, C are significant model terms.

7.12. Consider the putting experiment in Problem 6.21. Analyze the data considering each replicate as a block.

The analysis is similar to that of Problem 6.19. Blocking has not changed the significant factors, however, the residual plots show that the normality assumption has been violated. The transformed data also has similar analysis to the transformed data of Problem 6.21. The ANOVA shown is for the transformed data.

Design-Expert Output

Response: Distance from cup   Transform: Square root   Constant: 0
ANOVA for Selected Factorial Model
Analysis of variance table [Partial sum of squares]

Source	Sum of Squares	DF	Mean Square	F Value	Prob > F	
Block	13.50	6	2.25			
Model	37.26	2	18.63	7.83	0.0007	significant
A	21.61	1	21.61	9.08	0.0033	
B	15.64	1	15.64	6.57	0.0118	
Residual	245.13	103	2.38			
Cor Total	295.89	111				

The Model F-value of 7.83 implies the model is significant. There is only
a 0.07% chance that a "Model F-Value" this large could occur due to noise.

Values of "Prob > F" less than 0.0500 indicate model terms are significant.
In this case A, B are significant model terms.

7.22. Consider the 2^2 design in two blocks with AB confounded. Prove algebraically that $SS_{AB} = SS_{\text{Blocks}}$.

If AB is confounded, the two blocks are:

Block 1	Block 2
(1)	a
ab	b
$(1) + ab$	$a + b$

$$SS_{\text{Blocks}} = \frac{\left[(1) + ab\right]^2 + \left[a + b\right]^2}{2} - \frac{\left[(1) + ab + a + b\right]^2}{4}$$

$$SS_{\text{Blocks}} = \frac{(1)^2 + (ab)^2 + 2(1)ab + a^2 + b^2 + 2ab}{2}$$

$$- \frac{(1)^2 + (ab)^2 + a^2 + b^2 + 2(1)ab + 2(1)a + 2(1)b + 2a(ab) + 2b(ab) + 2ab}{4}$$

$$SS_{\text{Blocks}} = \frac{(1)^2 + (ab)^2 + a^2 + b^2 + 2(1)ab + 2ab - 2(1)a - 2(1)b - 2a(ab) - 2b(ab)}{4}$$

$$SS_{\text{Blocks}} = \frac{1}{4}\left[(1) + ab - a - b\right]^2 = SS_{AB}$$

CHAPTER 8

Two-Level Fractional Factorial Designs

LEARNING OBJECTIVES

After completing this chapter, you will be able to:

1. Plan, conduct, analyze, and interpret experiments involving a two-level fractional factorial design.

2. Understand the principles of constructing 2^{k-p} fractional factorial designs.

3. Understand how the alias structure of a 2^{k-p} fractional factorial design is determined.

4. Understand the concept of design resolution.

5. Explain how fractional factorial experiments work in terms of the sparsity of effects principle, design projection, and sequential experimentation.

6. Understand how fractional factorials can be combined sequentially to de-alias effects of interest.

KEY CONCEPTS AND IDEAS

1. Sparsity of effects
2. Design projection
3. Design generator
4. Defining relation
5. Alias

6. Design resolution
7. Sequential experimentation
8. Saturated design
9. Fold-over
10. Plackett-Burman design

Exercises

8.3. Consider the plasma etch experiment described in Example 6.1. Suppose that only a one-half fraction of the design could be run. Set up the design and analyze the data.

Because Example 6.1 is a replicated 2^3 factorial experiment, a half fraction of this design is a 2^{3-1} with four runs. The experiment is replicated to assure an adequate estimate of the MS_E.

A	B	C=AB	Etch Rate (A/min)		Factor Low (-)	Levels High (+)
-	-	+	1037	A (Gap, cm)	0.80	1.20
-	-	+	1052	B (C$_2$F$_6$ flow, SCCM)	125	200
+	-	-	669	C (Power, W)	275	325
+	-	-	650			
-	+	-	633			
-	+	-	601			
+	+	+	729			
+	+	+	860			

The analysis shown below identifies all three main effects as significant. Because this is a resolution III design, the main effects are aliased with two factor interactions. The original analysis from Example 6.1 identifies factors A, C, and the AC interaction as significant. In our replicated half fraction experiment, factor B is aliased with the AC interaction. This problem points out the concerns of running small resolution III designs.

Design-Expert Output

Response: Etch Rate						
ANOVA for Selected Factorial Model						
Analysis of variance table [Partial sum of squares]						
Source	Sum of Squares	DF	Mean Square	F Value	Prob > F	
Model	2.225E+005	3	74169.79	31.61	0.0030	significant
A	21528.13	1	21528.13	9.18	0.0388	
B	42778.13	1	42778.13	18.23	0.0130	
C	1.582E+005	1	1.582E+005	67.42	0.0012	
Pure Error	9385.50	4	2346.37			
Cor Total	2.319E+005	7				

The Model F-value of 31.61 implies the model is significant. There is only a 0.30% chance that a "Model F-Value" this large could occur due to noise.

Std. Dev.	48.44		R-Squared	0.9595	
Mean	778.88		Adj R-Squared	0.9292	
C.V.	6.22		Pred R-Squared	0.8381	
PRESS	37542.00		Adeq Precision	12.481	

Factor	Coefficient Estimate	DF	Standard Error	95% CI Low	95% CI High	VIF
Intercept	778.88	1	17.13	731.33	826.42	
A-Gap	-51.88	1	17.13	-99.42	-4.33	1.00
B-C2F6 Flow	-73.13	1	17.13	-120.67	-25.58	1.00
C-Power	140.63	1	17.13	93.08	188.17	1.00

```
Final Equation in Terms of Coded Factors:

       Etch Rate  =
         +778.88
          -51.88    * A
          -73.13    * B
         +140.63    * C

Final Equation in Terms of Actual Factors:

       Etch Rate  =
       -332.37500
       -259.37500   * Gap
         -1.95000   * C2F6 Flow
         +5.62500   * Power
```

8.4. Problem 6.26 describes a process improvement study in the manufacturing process of an integrated circuit. Suppose that only eight runs could be made in this process. Set up an appropriate 2^{5-2} design and find the alias structure. Use the appropriate observations from Problem 6.26 as the observations in this design and estimate the factor effects. What conclusions can you draw?

$$I = ABD = ACE = BCDE$$

A	(ABD)	=BD	A	(ACE)	=CE	A	(BCDE)	=ABCDE	A=BD=CE=ABCDE
B	(ABD)	=AD	B	(ACE)	=ABCE	B	(BCDE)	=CDE	B=AD=ABCE=CDE
C	(ABD)	=ABCD	C	(ACE)	=AE	C	(BCDE)	=BDE	C=ABCD=AE=BDE
D	(ABD)	=AB	D	(ACE)	=ACDE	D	(BCDE)	=BCE	D=AB=ACDE=BCE
E	(ABD)	=ABDE	E	(ACE)	=AC	E	(BCDE)	=BCD	E=ABDE=AC=BCD
BC	(ABD)	=ACD	BC	(ACE)	=ABE	BC	(BCDE)	=DE	BC=ACD=ABE=DE
BE	(ABD)	=ADE	BE	(ACE)	=ABC	BE	(BCDE)	=CD	BE=ADE=ABC=CD

A	B	C	D=AB	E=AC		
-	-	-	+	+	de	6
+	-	-	-	-	a	9
-	+	-	-	+	be	35
+	+	-	+	-	abd	50
-	-	+	+	-	cd	18
+	-	+	-	+	ace	22
-	+	+	-	-	bc	40
+	+	+	+	+	abcde	63

Design-Expert Output

	Term	Effect	SumSqr	% Contribtn
Model	Intercept			
Model	A	11.25	253.125	8.91953
Model	B	33.25	2211.13	77.9148
Model	C	10.75	231.125	8.1443
Model	D	7.75	120.125	4.23292
Error	E	2.25	10.125	0.356781
Error	BC	-1.75	6.125	0.215831
Error	BE	1.75	6.125	0.215831
	Lenth's ME	28.232		
	Lenth's SME	67.5646		

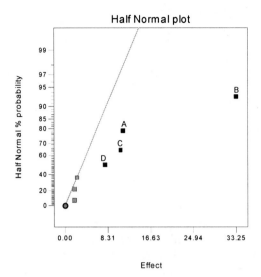

The main A, B, C, and D are large. However, recall that we are really estimating $A+BD+CE$, $B+AD$, $C+DE$ and $D+AD$. There are other possible interpretations of the experiment because of the aliasing.

Design-Expert Output

Response: Yield
ANOVA for Selected Factorial Model
Analysis of variance table [Partial sum of squares]

Source	Sum of Squares	DF	Mean Square	F Value	Prob > F	
Model	2815.50	4	703.88	94.37	0.0017	significant
A	*253.13*	*1*	*253.13*	*33.94*	*0.0101*	
B	*2211.12*	*1*	*2211.12*	*296.46*	*0.0004*	
C	*231.13*	*1*	*231.13*	*30.99*	*0.0114*	
D	*120.13*	*1*	*120.13*	*16.11*	*0.0278*	
Residual	22.38	3	7.46			
Cor Total	2837.88	7				

The Model F-value of 94.37 implies the model is significant. There is only a 0.17% chance that a "Model F-Value" this large could occur due to noise.

Std. Dev.	2.73	R-Squared	0.9921	
Mean	30.38	Adj R-Squared	0.9816	
C.V.	8.99	Pred R-Squared	0.9439	
PRESS	159.11	Adeq Precision	25.590	

Factor	Coefficient Estimate	DF	Standard Error	95% CI Low	95% CI High	VIF
Intercept	30.38	1	0.97	27.30	33.45	
A-Aperture	5.63	1	0.97	2.55	8.70	1.00
B-Exposure Time	16.63	1	0.97	13.55	19.70	1.00
C-Develop Time	5.37	1	0.97	2.30	8.45	1.00
D-Mask Dimension	3.87	1	0.97	0.80	6.95	1.00

Final Equation in Terms of Coded Factors:

Yield	=	
+30.38		
+5.63	* A	
+16.63	* B	
+5.37	* C	
+3.87	* D	

Final Equation in Terms of Actual Factors:

Aperture	Small
Mask Dimension	Small
Yield	=
-6.00000	
+0.83125	* Exposure Time
+0.71667	* Develop Time

Aperture	Large
Mask Dimension	Small
Yield	=
+5.25000	
+0.83125	* Exposure Time
+0.71667	* Develop Time

Aperture	Small
Mask Dimension	Large
Yield	=
+1.75000	
+0.83125	* Exposure Time
+0.71667	* Develop Time

Aperture	Large
Mask Dimension	Large
Yield	=
+13.00000	
+0.83125	* Exposure Time
+0.71667	* Develop Time

8.5. Continuation of Problem 8.4. Suppose you have made the eight runs in the 2^{5-2} design in Problem 8.4. What additional runs would be required to identify the factor effects that are of interest? What are the alias relationships in the combined design?

We could fold over the original design by changing the signs on the generators $D = AB$ and $E = AC$ to produce the following new experiment.

A	B	C	D=-AB	E=-AC		
-	-	-	-	-	(1)	7
+	-	-	+	+	ade	10
-	+	-	+	-	bd	32
+	+	-	-	+	abe	52
-	-	+	-	+	ce	15
+	-	+	+	-	acd	21
-	+	+	+	+	bcde	41
+	+	+	-	-	abc	60

A	(-ABD)	=-BD	A	(-ACE)	=-CE	A	(BCDE)	=ABCDE	A=-BD=-CE=ABCDE
B	(-ABD)	=-AD	B	(-ACE)	=-ABCE	B	(BCDE)	=CDE	B=-AD=-ABCE=CDE
C	(-ABD)	=-ABCD	C	(-ACE)	=-AE	C	(BCDE)	=BDE	C=-ABCD=-AE=BDE
D	(-ABD)	=-AB	D	(-ACE)	=-ACDE	D	(BCDE)	=BCE	D=-AB=-ACDE=BCE
E	(-ABD)	=-ABDE	E	(-ACE)	=-AC	E	(BCDE)	=BCD	E=-ABDE=-AC=BCD
BC	(-ABD)	=-ACD	BC	(-ACE)	=-ABE	BC	(BCDE)	=DE	BC=-ACD=-ABE=DE
BE	(-ABD)	=-ADE	BE	(-ACE)	=-ABC	BE	(BCDE)	=CD	BE=-ADE=-ABC=CD

Assuming all three factor and higher interactions to be negligible, all main effects can be separated from their two-factor interaction aliases in the combined design.

8.9. R.D. Snee ("Experimenting with a Large Number of Variables," in *Experiments in Industry: Design, Analysis and Interpretation of Results*, by R.D. Snee, L.B. Hare, and J.B. Trout, Editors, ASQC, 1985) describes an experiment in which a 2^{5-1} design with $I=ABCDE$ was used to investigate the effects of five factors on the color of a chemical product. The factors are A = solvent/reactant, B = catalyst/reactant, C = temperature, D = reactant purity, and E = reactant pH. The results obtained were as follows:

$e =$	-0.63	$d =$	6.79
$a =$	2.51	$ade =$	5.47
$b =$	-2.68	$bde =$	3.45
$abe =$	1.66	$abd =$	5.68
$c =$	2.06	$cde =$	5.22
$ace =$	1.22	$acd =$	4.38
$bce =$	-2.09	$bcd =$	4.30
$abc =$	1.93	$abcde =$	4.05

(a) Prepare a normal probability plot of the effects. Which effects seem active?

Factors A, B, D, and the AB, AD interactions appear to be active.

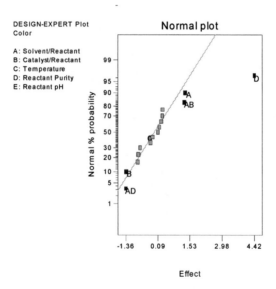

Design-Expert Output

	Term	Effect	SumSqr	% Contribtn
Model	Intercept			
Model	A	1.31	6.8644	5.98537
Model	B	-1.34	7.1824	6.26265
Error	C	-0.1475	0.087025	0.0758809
Model	D	4.42	78.1456	68.1386
Error	E	-0.8275	2.73902	2.38828
Model	AB	1.275	6.5025	5.66981
Error	AC	-0.7875	2.48062	2.16297
Model	AD	-1.355	7.3441	6.40364
Error	AE	0.3025	0.366025	0.319153
Error	BC	0.1675	0.112225	0.0978539
Error	BD	0.245	0.2401	0.209354
Error	BE	0.2875	0.330625	0.288286
Error	CD	-0.7125	2.03063	1.77059
Error	CE	-0.24	0.2304	0.200896
Error	DE	0.0875	0.030625	0.0267033
	Lenth's ME	1.95686		
	Lenth's SME	3.9727		

Design-Expert Output

Response: Color
ANOVA for Selected Factorial Model
Analysis of variance table [Partial sum of squares]

Source	Sum of Squares	DF	Mean Square	F Value	Prob > F	
Model	106.04	5	21.21	24.53	< 0.0001	significant
A	6.86	1	6.86	7.94	0.0182	
B	7.18	1	7.18	8.31	0.0163	
D	78.15	1	78.15	90.37	< 0.0001	
AB	6.50	1	6.50	7.52	0.0208	
AD	7.34	1	7.34	8.49	0.0155	
Residual	8.65	10	0.86			
Cor Total	114.69	15				

The Model F-value of 24.53 implies the model is significant. There is only
a 0.01% chance that a "Model F-Value" this large could occur due to noise.

Std. Dev.	0.93	R-Squared	0.9246	
Mean	2.71	Adj R-Squared	0.8869	
C.V.	34.35	Pred R-Squared	0.8070	
PRESS	22.14	Adeq Precision	14.734	

Factor	Coefficient Estimate	DF	Standard Error	95% CI Low	95% CI High	VIF
Intercept	2.71	1	0.23	2.19	3.23	
A-Solvent/Reactant	0.66	1	0.23	0.14	1.17	1.00
B-Catalyst/Reactant	-0.67	1	0.23	-1.19	-0.15	1.00
D-Reactant Purity	2.21	1	0.23	1.69	2.73	1.00
AB	0.64	1	0.23	0.12	1.16	1.00
AD	-0.68	1	0.23	-1.20	-0.16	1.00

Final Equation in Terms of Coded Factors:

$$Color = $$
$$+2.71$$
$$+0.66 * A$$
$$-0.67 * B$$
$$+2.21 * D$$
$$+0.64 * A * B$$
$$-0.68 * A * D$$

Final Equation in Terms of Actual Factors:

Color   =
+2.70750
+0.65500   * Solvent/Reactant
-0.67000   * Catalyst/Reactant
+2.21000   * Reactant Purity
+0.63750   * Solvent/Reactant * Catalyst/Reactant
-0.67750   * Solvent/Reactant * Reactant Purity

(b) Calculate the residuals. Construct a normal probability plot of the residuals and plot the residuals versus the fitted values. Comment on the plots.

Design-Expert Output

Diagnostics Case Statistics

Standard Order	Actual Value	Predicted Value	Residual	Leverage	Student Residual	Cook's Distance	Outlier t	Run Order
1	-0.63	0.47	-1.10	0.375	-1.500	0.225	-1.616	2
2	2.51	1.86	0.65	0.375	0.881	0.078	0.870	6
3	-2.68	-2.14	-0.54	0.375	-0.731	0.053	-0.713	14
4	1.66	1.80	-0.14	0.375	-0.187	0.003	-0.178	11
5	2.06	0.47	1.59	0.375	2.159	0.466	2.804	8
6	1.22	1.86	-0.64	0.375	-0.874	0.076	-0.863	15
7	-2.09	-2.14	0.053	0.375	0.071	0.001	0.068	10
8	1.93	1.80	0.13	0.375	0.180	0.003	0.171	3
9	6.79	6.25	0.54	0.375	0.738	0.054	0.720	4
10	5.47	4.93	0.54	0.375	0.738	0.054	0.720	5
11	3.45	3.63	-0.18	0.375	-0.248	0.006	-0.236	16
12	5.68	4.86	0.82	0.375	1.112	0.124	1.127	12
13	5.22	6.25	-1.03	0.375	-1.398	0.195	-1.478	9
14	4.38	4.93	-0.55	0.375	-0.745	0.055	-0.727	1
15	4.30	3.63	0.67	0.375	0.908	0.082	0.899	13
16	4.05	4.86	-0.81	0.375	-1.105	0.122	-1.119	7

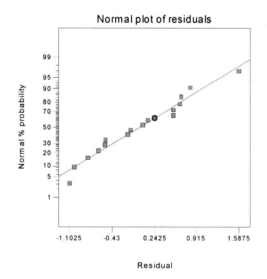

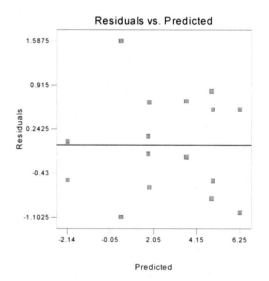

The residual plots are satisfactory.

(c) If any factors are negligible, collapse the 2^{5-1} design into a full factorial in the active factors. Comment on the resulting design, and interpret the results.

The design becomes two replicates of a 2^3 in the factors A, B and D. When re-analyzing the data in three factors, D becomes labeled as C.

Design-Expert Output

Response: Color
    ANOVA for Selected Factorial Model
Analysis of variance table [Partial sum of squares]

Source	Sum of Squares	DF	Mean Square	F Value	Prob > F	
Model	106.51	7	15.22	14.89	0.0005	significant
A	*6.86*	*1*	*6.86*	*6.72*	*0.0320*	
B	*7.18*	*1*	*7.18*	*7.03*	*0.0292*	
C	*78.15*	*1*	*78.15*	*76.46*	*< 0.0001*	
AB	*6.50*	*1*	*6.50*	*6.36*	*0.0357*	
AC	*7.34*	*1*	*7.34*	*7.19*	*0.0279*	
BC	*0.24*	*1*	*0.24*	*0.23*	*0.6409*	
ABC	*0.23*	*1*	*0.23*	*0.23*	*0.6476*	
Residual	8.18	8	1.02			
Lack of Fit	*0.000*	*0*				
Pure Error	*8.18*	*8*	*1.02*			
Cor Total	114.69	15				

The Model F-value of 14.89 implies the model is significant. There is only a 0.05% chance that a "Model F-Value" this large could occur due to noise.

Std. Dev.	1.01	R-Squared	0.9287	
Mean	2.71	Adj R-Squared	0.8663	
C.V.	37.34	Pred R-Squared	0.7148	
PRESS	32.71	Adeq Precision	11.736	

Factor	Coefficient Estimate	DF	Standard Error	95% CI Low	95% CI High	VIF
Intercept	2.71	1	0.25	2.12	3.29	
A-Solvent/Reactant	0.66	1	0.25	0.072	1.24	1.00
B-Catalyst/Reactant	-0.67	1	0.25	-1.25	-0.087	1.00
C-Reactant Purity	2.21	1	0.25	1.63	2.79	1.00
AB	0.64	1	0.25	0.055	1.22	1.00
AC	-0.68	1	0.25	-1.26	-0.095	1.00
BC	0.12	1	0.25	-0.46	0.71	1.00
ABC	-0.12	1	0.25	-0.70	0.46	1.00

Final Equation in Terms of Coded Factors:

Color	=	
+2.71		
+0.66	* A	
-0.67	* B	
+2.21	* C	
+0.64	* A * B	
-0.68	* A * C	
+0.12	* B * C	
-0.12	* A * B * C	

Final Equation in Terms of Actual Factors:

Color =
+2.70750
+0.65500 * Solvent/Reactant
-0.67000 * Catalyst/Reactant
+2.21000 * Reactant Purity
+0.63750 * Solvent/Reactant * Catalyst/Reactant
-0.67750 * Solvent/Reactant * Reactant Purity
+0.12250 * Catalyst/Reactant * Reactant Purity
-0.12000 * Solvent/Reactant * Catalyst/Reactant * Reactant Purity

8.27. Carbon anodes used in a smelting process are baked in a ring furnace. An experiment is run in the furnace to determine which factors influence the weight of packing material that is stuck to the anodes after baking. Six variables are of interest, each at two levels: A = pitch/fines ratio (0.45, 0.55); B = packing material type (1, 2); C = packing material temperature (ambient, 325 °C); D = flue location (inside, outside); E = pit temperature (ambient, 195 °C); and F = delay time before packing (zero, 24 hours). A 2^{6-3} design is run, and three replicates are obtained at each of the design points. The weight of packing material stuck to the anodes is measured in grams. The data in run order are as follows: abd = (984, 826, 936); $abcdef$ = (1275, 976, 1457); be = (1217, 1201, 890); af = (1474, 1164, 1541); def = (1320, 1156, 913); cd = (765, 705, 821); ace = (1338, 1254, 1294); and bcf = (1325, 1299, 1253). We wish to minimize the amount of stuck packing material.

(a) Verify that the eight runs correspond to a 2_{III}^{6-3} design. What is the alias structure?

A	B	C	D=AB	E=AC	F=BC	
-	-	-	+	+	+	def
+	-	-	-	-	+	af
-	+	-	-	+	-	be
+	+	-	+	-	-	abd
-	-	+	+	-	-	cd
+	-	+	-	+	-	ace
-	+	+	-	-	+	bcf
+	+	+	+	+	+	abcdef

I=ABD=ACE=BCF=BCDE=ACDF=ABEF=DEF, Resolution III

A=BD=CE=CDF=BEF
B=AD=CF=CDE=AEF
C=AE=BF=BDE=ADF
D=AB=EF=BCE=ACF
E=AC=DF=BCD=ABF
F=BC=DE=ACD=ABE
CD=BE=AF=ABC=ADE=BDF=CEF

(b) Use the average weight as a response. What factors appear to be influential?

Design-Expert Output

	Term	Effect	SumSqr	% Contribtn
Model	A	137.833	37996.1	12.0947
Error	B	-8.83333	156.056	0.049675
Error	C	11.6667	272.222	0.0866527
Model	D	-259.667	134854	42.926
Model	E	99.8333	19933.4	6.34511
Model	F	243.5	118585	37.7473
Error	AF	-34.3333	2357.56	0.750447
	Lenth's ME	563.698		
	Lenth's SME	1349.04		

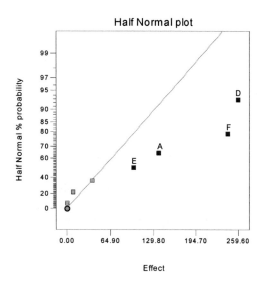

Half Normal plot

Factors *A, D, E* and *F* (and their aliases) are apparently important.

(c) Use the range of the weights as a response. What factors appear to be influential?

Design-Expert Output

	Term	Effect	SumSqr	% Contribtn
Model	Intercept			
Error	A	44.5	3960.5	2.13311
Error	B	13.5	364.5	0.196319
Model	C	-129	33282	17.9256
Error	D	75.5	11400.5	6.14028
Model	E	144	41472	22.3367
Model	F	163	53138	28.62
Model	AF	145	42050	22.648
	Lenth's ME	728.384		
	Lenth's SME	1743.17		

Factors *C, E, F* and the *AF* interaction (and their aliases) appear to be large.

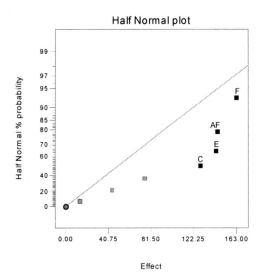

Half Normal plot

(d) What recommendations would you make to the process engineers?

It is not known exactly what to do here, since *A, D, E* and *F* are large effects, and because the design is resolution III, the main effects are aliased with two-factor interactions. Note, for example, that *D* is aliased with *EF* and the main effect could really be an *EF* interaction. If the main effects are really important, then setting all factors at the low level would minimize the amount of material stuck to the anodes. It would be necessary to run additional experiments to confirm these findings.

8.28. A 16-run experiment was performed in a semiconductor manufacturing plant to study the effects of six factors on the curvature or camber of the substrate devices produced. The six variables and their levels are shown in Table P8.2.

Table P8.2

Run	Lamination Temperature (c)	Lamination Time (s)	Lamination Pressure (tn)	Firing Temperature (c)	Firing Cycle Time (h)	Firing Dew Point (c)
1	55	10	5	1580	17.5	20
2	75	10	5	1580	29	26
3	55	25	5	1580	29	20
4	75	25	5	1580	17.5	26
5	55	10	10	1580	29	26
6	75	10	10	1580	17.5	20
7	55	25	10	1580	17.5	26
8	75	25	10	1580	29	20
9	55	10	5	1620	17.5	26
10	75	10	5	1620	29	20
11	55	25	5	1620	29	26
12	75	25	5	1620	17.5	20
13	55	10	10	1620	29	20
14	75	10	10	1620	17.5	26
15	55	25	10	1620	17.5	20
16	75	25	10	1620	29	26

Each run was replicated four times, and a camber measurement was taken on the substrate. The data are shown in the following table.

	Camber	for	Replicate	(in/in)	Total	Mean	Standard
Run	1	2	3	4	$(10^{-4}$ in/in)	$(10^{-4}$ in/in)	Deviation
1	0.0167	0.0128	0.0149	0.0185	629	157.25	24.418
2	0.0062	0.0066	0.0044	0.0020	192	48.00	20.976
3	0.0041	0.0043	0.0042	0.0050	176	44.00	4.083
4	0.0073	0.0081	0.0039	0.0030	223	55.75	25.025
5	0.0047	0.0047	0.0040	0.0089	223	55.75	22.410
6	0.0219	0.0258	0.0147	0.0296	920	230.00	63.639
7	0.0121	0.0090	0.0092	0.0086	389	97.25	16.029
8	0.0255	0.0250	0.0226	0.0169	900	225.00	39.420
9	0.0032	0.0023	0.0077	0.0069	201	50.25	26.725
10	0.0078	0.0158	0.0060	0.0045	341	85.25	50.341
11	0.0043	0.0027	0.0028	0.0028	126	31.50	7.681
12	0.0186	0.0137	0.0158	0.0159	640	160.00	20.083
13	0.0110	0.0086	0.0101	0.0158	455	113.75	31.120
14	0.0065	0.0109	0.0126	0.0071	371	92.75	29.510
15	0.0155	0.0158	0.0145	0.0145	603	150.75	6.750
16	0.0093	0.0124	0.0110	0.0133	460	115.00	17.450

(a) What type of design did the experimenters use?

The 2_{II}^{6-2}, a 16-run design.

(b) What are the alias relationships in this design?

The defining relation is I=$ABCE$=$ACDF$=$BDEF$. The aliases are shown below.

$A(ABCE)=$	BCE	$A(ACDF)=$	CDF	$A(BDEF)=$	$ABDEF$	$A=BCE=CDF=ABDEF$
$B(ABCE)=$	ACE	$B(ACDF)=$	$ABCDF$	$B(BDEF)=$	DEF	$B=ACE=ABCDF=DEF$
$C(ABCE)=$	ABE	$C(ACDF)=$	ADF	$C(BDEF)=$	$BCDEF$	$C=ABE=ADF=BCDEF$
$D(ABCE)=$	$ABCDE$	$D(ACDF)=$	ACF	$D(BDEF)=$	BEF	$D=ABCDE=ACF=BEF$
$E(ABCE)=$	ABC	$E(ACDF)=$	$ACDEF$	$E(BDEF)=$	BDF	$E=ABC=ACDEF=BDF$
$F(ABCE)=$	$ABCEF$	$F(ACDF)=$	ACD	$F(BDEF)=$	BDE	$F=ABCEF=ACD=BDE$
$AB(ABCE)=$	CE	$AB(ACDF)=$	$BCDF$	$AB(BDEF)=$	$ADEF$	$AB=CE=BCDF=ADEF$
$AC(ABCE)=$	BE	$AC(ACDF)=$	DF	$AC(BDEF)=$	$ABCDEF$	$AC=BE=DF=ABCDEF$
$AD(ABCE)=$	$BCDE$	$AD(ACDF)=$	CF	$AD(BDEF)=$	$ABEF$	$AD=BCDE=CF=ABEF$
$AE(ABCE)=$	BC	$AE(ACDF)=$	$CDEF$	$AE(BDEF)=$	$ABDF$	$AE=BC=CDEF=ABDF$
$AF(ABCE)=$	$BCEF$	$AF(ACDF)=$	CD	$AF(BDEF)=$	$ABDE$	$AF=BCEF=CD=ABDE$
$BD(ABCE)=$	$ACDE$	$BD(ACDF)=$	$ABCF$	$BD(BDEF)=$	EF	$BD=ACDE=ABCF=EF$
$BF(ABCE)=$	$ACEF$	$BF(ACDF)=$	$ABCD$	$BF(BDEF)=$	DE	$BF=ACEF=ABCD=DE$

(c) Do any of the process variables affect average camber?

Yes, per the analysis below, variables A, C, E, and F affect average camber.

Design-Expert Output

	Term	Effect	SumSqr	% Contribtn
Model	Intercept			
Model	A	38.9063	6054.79	10.2962
Error	B	5.78125	133.691	0.227344
Model	C	56.0313	12558	21.355
Error	D	-14.2188	808.691	1.37519
Model	E	-34.4687	4752.38	8.08148
Model	F	-77.4688	24005.6	40.8219
Error	AB	19.1563	1467.85	2.49609
Error	AC	22.4063	2008.16	3.4149
Error	AD	-12.2188	597.191	1.01553
Error	AE	18.1563	1318.6	2.24229

Error	AF	-19.7187	1555.32	2.64483
Error	BC	Aliased		
Error	BD	23.0313	2121.75	3.60807
Error	BE	Aliased		
Error	BF	7.40625	219.41	0.37311
Error	CD	Aliased		
Error	CE	Aliased		
Error	CF	Aliased		
Error	DE	Aliased		
Error	DF	Aliased		
Error	EF	Aliased		
Error	ABC	Aliased		
Error	ABD	0.53125	1.12891	0.00191972
Error	ABE	Aliased		
Error	ABF	-17.3438	1203.22	2.04609
	Lenth's ME	71.9361		
	Lenth's SME	146.041		

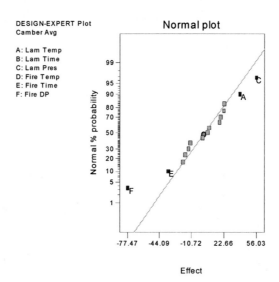

Design-Expert Output

Response: Camber Avg in in/in
  ANOVA for Selected Factorial Model
Analysis of variance table [Partial sum of squares]

Source	Sum of Squares	DF	Mean Square	F Value	Prob > F	
Model	47370.80	4	11842.70	11.39	0.0007	significant
A	6054.79	1	6054.79	5.82	0.0344	
C	12558.00	1	12558.00	12.08	0.0052	
E	4752.38	1	4752.38	4.57	0.0558	
F	24005.63	1	24005.63	23.09	0.0005	
Residual	11435.01	11	1039.55			
Cor Total	58805.81	15				

The Model F-value of 11.39 implies the model is significant. There is only a 0.07% chance that a "Model F-Value" this large could occur due to noise.

Std. Dev.	32.24	R-Squared	0.8055	
Mean	107.02	Adj R-Squared	0.7348	
C.V.	30.13	Pred R-Squared	0.5886	
PRESS	24193.08	Adeq Precision	11.478	

Factor	Coefficient Estimate	DF	Standard Error	95% CI Low	95% CI High	VIF
Intercept	107.02	1	8.06	89.27	124.76	
A-Lam Temp	19.45	1	8.06	1.71	37.19	1.00
C-Lam Pres	28.02	1	8.06	10.27	45.76	1.00
E-Fire Time	-17.23	1	8.06	-34.98	0.51	1.00
F-Fire DP	-38.73	1	8.06	-56.48	-20.99	1.00

Final Equation in Terms of Coded Factors:

Camber Avg =
+107.02
+19.45   * A
+28.02   * C
-17.23   * E
-38.73   * F

Final Equation in Terms of Actual Factors:

Camber Avg =
+263.17380
+1.94531   * Lam Temp
+11.20625   * Lam Pres
-2.99728   * Fire Time
-12.91146   * Fire DP

(d) Do any of the process variables affect the variability in camber measurements?

Yes, A, B, F, and AF interaction affect the variability in camber measurements.

Design-Expert Output

	Term	Effect	SumSqr	% Contribtn
Model	Intercept			
Model	A	15.9035	1011.69	27.6623
Model	B	-16.5773	1099.22	30.0558
Error	C	5.8745	138.039	3.77437
Error	D	-3.2925	43.3622	1.18564
Error	E	-2.33725	21.851	0.597466
Model	F	-9.256	342.694	9.37021
Error	AB	0.95525	3.65001	0.0998014
Error	AC	2.524	25.4823	0.696757
Error	AD	-4.6265	85.618	2.34103
Error	AE	-0.18025	0.12996	0.00355347
Model	AF	-10.8745	473.019	12.9337
Error	BC	Aliased		
Error	BD	-4.85575	94.3132	2.57879
Error	BE	Aliased		
Error	BF	8.21825	270.159	7.38689
Error	CD	Aliased		
Error	CE	Aliased		
Error	CF	Aliased		
Error	DE	Aliased		
Error	DF	Aliased		
Error	EF	Aliased		
Error	ABC	Aliased		
Error	ABD	-0.68125	1.85641	0.0507593
Error	ABE	Aliased		
Error	ABF	3.39825	46.1924	1.26303
	Lenth's ME	17.8392		
	Lenth's SME	36.2162		

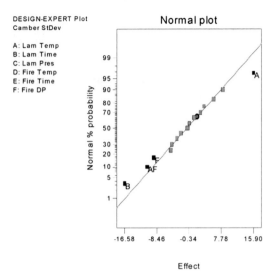

DESIGN-EXPERT Plot
Camber StDev

A: Lam Temp
B: Lam Time
C: Lam Pres
D: Fire Temp
E: Fire Time
F: Fire DP

Response: Camber StDev
    ANOVA for Selected Factorial Model
Analysis of variance table [Partial sum of squares]

Source	Sum of Squares	DF	Mean Square	F Value	Prob > F	
Model	2926.62	4	731.65	11.02	0.0008	significant
A	1011.69	1	1011.69	15.23	0.0025	
B	1099.22	1	1099.22	16.55	0.0019	
F	342.69	1	342.69	5.16	0.0442	
AF	473.02	1	473.02	7.12	0.0218	
Residual	730.65	11	66.42			
Cor Total	3657.27	15				

The Model F-value of 11.02 implies the model is significant. There is only
a 0.08% chance that a "Model F-Value" this large could occur due to noise.

Std. Dev.	8.15	R-Squared	0.8002	
Mean	25.35	Adj R-Squared	0.7276	
C.V.	32.15	Pred R-Squared	0.5773	
PRESS	1545.84	Adeq Precision	9.516	

Factor	Coefficient Estimate	DF	Standard Error	95% CI Low	95% CI High	VIF
Intercept	25.35	1	2.04	20.87	29.84	
A-Lam Temp	7.95	1	2.04	3.47	12.44	1.00
B-Lam Time	-8.29	1	2.04	-12.77	-3.80	1.00
F-Fire DP	-4.63	1	2.04	-9.11	-0.14	1.00
AF	-5.44	1	2.04	-9.92	-0.95	1.00

Final Equation in Terms of Coded Factors:

Camber StDev   =
    +25.35
    +7.95   * A
    -8.29   * B
    -4.63   * F
    -5.44   * A * F

Final Equation in Terms of Actual Factors:

Camber StDev   =
-242.46746
    +4.96373   * Lam Temp
    -1.10515   * Lam Time
    +10.23804   * Fire DP
    -0.18124   * Lam Temp * Fire DP

(e) If it is important to reduce camber as much as possible, what recommendations would you make?

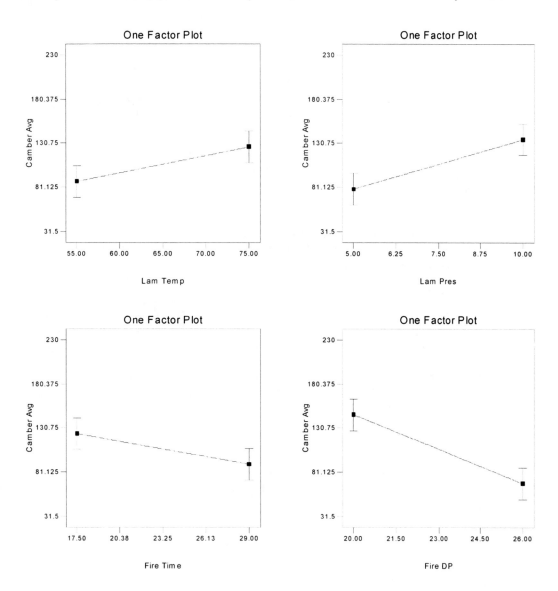

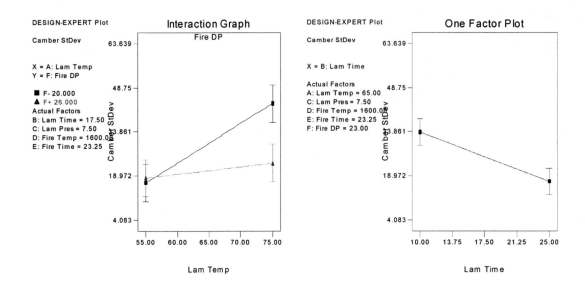

Run A and C at the low level and E and F at the high level. B at the low level enables a lower variation without affecting the average camber.

8.30. Harry Peterson-Nedry (a friend of the author) own a vineyard and winery in Newberg, Oregon. He grows several varieties of grapes and produces wine. Harry has used factorial designs for process and product development in the winemaking segment of their business. This problem describes the experiment conducted for their 1985 Pinot Noir. Eight variables, shown in Table P8.5, were originally studied in this experiment:

Table P8.5

	Variable	Low Level	High Level
A	Pinot Noir Clone	Pommard	Wadenswil
B	Oak Type	Allier	Troncais
C	Age of Barrel	Old	New
D	Yeast/Skin Contact	Champagne	Montrachet
E	Stems	None	All
F	Barrel Toast	Light	Medium
G	Whole Cluster	None	10%
H	Fermentation Temperature	Low (75 F Max)	High (92 F Max)

Harry and Judy decided to use a 2_{IV}^{8-4} design with 16 runs. The wine was taste-tested by a panel of experts on March 8, 1986. Each expert ranked the 16 samples of wine tasted, with rank 1 being the best. The design and taste-test panel results are shown in Table P8.4.

Table P8.6

Run	A	B	C	D	E	F	G	H	HPN	JPN	CAL	DCM	RGB	$\bar{y}_{bar}$	s
1	-	-	-	-	-	-	-	-	12	6	13	10	7	9.6	3.05
2	+	-	-	-	-	+	+	+	10	7	14	14	9	10.8	3.11
3	-	+	-	-	+	-	+	+	14	13	10	11	15	12.6	2.07
4	+	+	-	-	+	+	-	-	9	9	7	9	12	9.2	1.79
5	-	-	+	-	+	+	+	-	8	8	11	8	10	9.0	1.41
6	+	-	+	-	+	-	-	+	16	12	15	16	16	15.0	1.73
7	-	+	+	-	-	+	-	+	6	5	6	5	3	5.0	1.22
8	+	+	+	-	-	-	+	-	15	16	16	15	14	15.2	0.84
9	-	-	-	+	+	+	-	+	1	2	3	3	2	2.2	0.84
10	+	-	-	+	+	-	+	-	7	11	4	7	6	7.0	2.55
11	-	+	-	+	-	+	+	-	13	3	8	12	8	8.8	3.96
12	+	+	-	+	-	-	-	+	3	1	5	1	4	2.8	1.79
13	-	-	+	+	-	-	+	+	2	10	2	4	5	4.6	3.29
14	+	-	+	+	-	+	-	-	4	4	1	2	1	2.4	1.52
15	-	+	+	+	+	-	-	-	5	15	9	6	11	9.2	4.02
16	+	+	+	+	+	+	+	+	11	14	12	13	13	12.6	1.14

(a) What are the alias relationships in the design selected by Harry and Judy?

$$E = BCD, \; F = ACD, \; G = ABC, \; H = ABD$$

Defining Contrast : $\; I = BCDE = ACDF = ABEF = ABCG = ADEG = BDFG = CEFG = ABDH$
$$= ACEH = BCFH = DEFH = CDGH = BEGH = AFGH = ABCDEFGH$$

Aliases:

$$A = BCG = BDH = BEF = CDF = CEH = DEG = FGH$$
$$B = ACG = ADH = AEF = CDE = CFH = DFG = EGH$$
$$C = ABG = ADF = AEH = BDE = BFH = DGH = EFG$$
$$D = ABH = ACF = AEG = BCE = BFG = CGH = EFH$$
$$E = ABF = ACH = ADG = BCD = BGH = CFG = DFH$$
$$F = ABE = ACD = AGH = BCH = BDG = CEG = DEH$$
$$G = ABC = ADE = AFH = BDF = BEH = CDH = CEF$$
$$H = ABD = ACE = AFG = BCF = BEG = CDG = DEF$$
$$AB = CG = DH = EF$$
$$AC = BG = DF = EH$$
$$AD = BH = CF = EG$$
$$AE = BF = CH = DG$$
$$AF = BE = CD = GH$$
$$AG = BC = DE = FH$$
$$AH = BD = CE = FG$$

(b) Use the average ranks ($\bar{y}$) as a response variable. Analyze the data and draw conclusions. You will find it helpful to examine a normal probability plot of effect estimates.

The effects list and normal probability plot of effects are shown below. Factors D, E, F, and G appear to be significant. Also note that the DF and FG interactions were chosen instead of AC and AH based on the alias structure shown above.

Design-Expert Output

	Term	Effect	SumSqr	% Contribtn
Require	Intercept			
Error	A	1.75	12.25	4.57636
Error	B	1.85	13.69	5.11432
Error	C	1.25	6.25	2.33488
Model	D	-4.6	84.64	31.6198
Model	E	2.2	19.36	7.23252
Model	F	-2	16	5.97729
Model	G	3.15	39.69	14.8274
Error	H	-0.6	1.44	0.537956
Error	AB	-0.7	1.96	0.732218
Ignore	AC	Aliased		
Error	AD	-1.75	12.25	4.57636
Error	AE	0.95	3.61	1.34863
Error	AF	0.75	2.25	0.840556
Error	AG	0.9	3.24	1.2104
Ignore	AH	Aliased		
Model	DF	2.6	27.04	10.1016
Model	FG	2.45	24.01	8.96967
	Lenth's ME	6.74778		
	Lenth's SME	13.699		

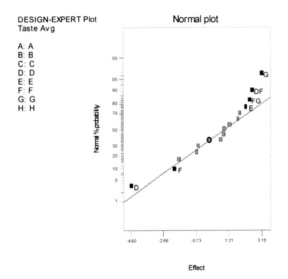

DESIGN-EXPERT Plot
Taste Avg

A: A
B: B
C: C
D: D
E: E
F: F
G: G
H: H

Normal plot

Normal % probability

Effect

Design-Expert Output

Response: Taste Avg
    ANOVA for Selected Factorial Model
Analysis of variance table [Partial sum of squares]

Source	Sum of Squares	DF	Mean Square	F Value	Prob > F	
Model	210.74	6	35.12	5.55	0.0115	significant
D	84.64	1	84.64	13.38	0.0053	
E	19.36	1	19.36	3.06	0.1142	
F	16.00	1	16.00	2.53	0.1462	
G	39.69	1	39.69	6.27	0.0336	
DF	27.04	1	27.04	4.27	0.0687	
FG	24.01	1	24.01	3.80	0.0832	
Residual	56.94	9	6.33			
Cor Total	267.68	15				

The Model F-value of 5.55 implies the model is significant. There is only a 1.15% chance that a "Model F-Value" this large could occur due to noise.

Std. Dev.	2.52	R-Squared	0.7873
Mean	8.50	Adj R-Squared	0.6455
C.V.	29.59	Pred R-Squared	0.3277
PRESS	179.96	Adeq Precision	7.183

Factor	Coefficient Estimate	DF	Standard Error	95% CI Low	95% CI High	VIF
Intercept	8.50	1	0.63	7.08	9.92	
D-D	-2.30	1	0.63	-3.72	-0.88	1.00
E-E	1.10	1	0.63	-0.32	2.52	1.00
F-F	-1.00	1	0.63	-2.42	0.42	1.00
G-G	1.57	1	0.63	0.15	3.00	1.00
DF	1.30	1	0.63	-0.12	2.72	1.00
FG	1.23	1	0.63	-0.20	2.65	1.00

Final Equation in Terms of Coded Factors:

$$\begin{aligned}
\text{Taste Avg} = \\
+8.50 \\
-2.30 \;\; * D \\
+1.10 \;\; * E \\
-1.00 \;\; * F \\
+1.57 \;\; * G \\
+1.30 \;\; * D * F \\
+1.23 \;\; * F * G
\end{aligned}$$

Factors D and G and are important. Factor E and the DF and FG interactions are moderately important and were included in the model because the PRESS statistic showed improvement with their inclusion. Factor F is added to the model to preserve hierarchy. As stated earlier, the interactions are aliased with other two-factor interactions that could also be important. So the interpretation of the two-factor interaction is somewhat uncertain. Normally, we would add runs to the design to isolate the significant interactions, but that will not work very well here because each experiment requires a full growing season. In other words, it would require a very long time to add runs to de-alias the alias chains of interest.

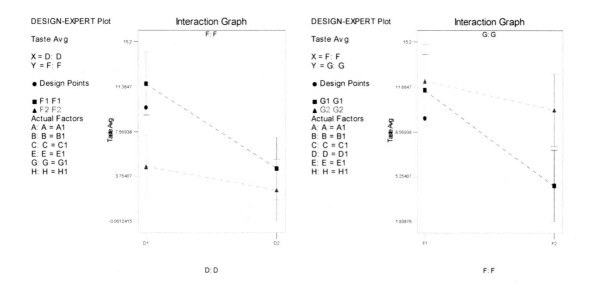

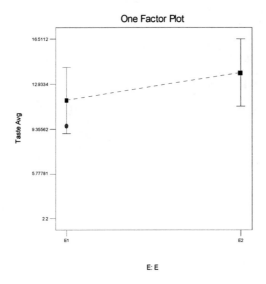

(c) Use the standard deviation of the ranks (or some appropriate transformation such as log *s*) as a response variable. What conclusions can you draw about the effects of the eight variables on variability in wine quality?

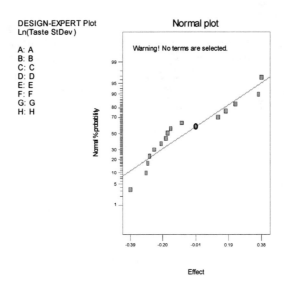

There do not appear to be any significant factors.

(d) After looking at the results, Harry and Judy decide that one of the panel members (DCM) knows more about beer than he does about wine, so they decide to delete his ranking. What effect would this have on the results and on conclusions from parts (b) and (c)?

Design-Expert Output

	Term	Effect	SumSqr	% Contribtn
Require	Intercept			
Error	A	1.625	10.5625	4.02957
Error	B	2.0625	17.0156	6.49142
Error	C	1.5	9	3.43348
Model	D	-4.5	81	30.9013
Model	E	2.4375	23.7656	9.06652
Model	F	-2.375	22.5625	8.60753
Model	G	2.9375	34.5156	13.1676
Error	H	-0.6875	1.89063	0.721268
Error	AB	-0.5625	1.26563	0.482833
Ignore	AC	Aliased		
Error	AD	-1.5	9	3.43348
Error	AE	0.6875	1.89063	0.721268
Error	AF	0.875	3.0625	1.16834
Error	AG	0.8125	2.64062	1.00739
Ignore	AH	Aliased		
Model	DF	2.375	22.5625	8.60753
Model	FG	2.3125	21.3906	8.16047
	Lenth's ME	6.26579		
	Lenth's SME	12.7205		

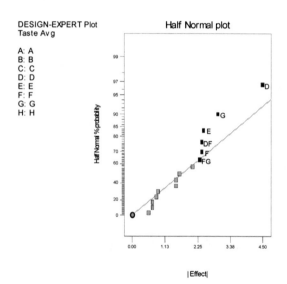

DESIGN-EXPERT Plot
Taste Avg

A: A
B: B
C: C
D: D
E: E
F: F
G: G
H: H

Half Normal plot

Design-Expert Output

Response: Taste Avg
ANOVA for Selected Factorial Model
Analysis of variance table [Partial sum of squares]

Source	Sum of Squares	DF	Mean Square	F Value	Prob > F	
Model	205.80	6	34.30	5.48	0.0120	significant
D	81.00	1	81.00	12.94	0.0058	
E	23.77	1	23.77	3.80	0.0831	
F	22.56	1	22.56	3.60	0.0901	
G	34.52	1	34.52	5.51	0.0434	
DF	22.56	1	22.56	3.60	0.0901	
FG	21.39	1	21.39	3.42	0.0975	
Residual	56.33	9	6.26			
Cor Total	262.13	15				

The Model F-value of 5.48 implies the model is significant. There is only a 1.20% chance that a "Model F-Value" this large could occur due to noise.

Std. Dev.	2.50		R-Squared	0.7851		
Mean	8.50		Adj R-Squared	0.6418		
C.V.	29.43		Pred R-Squared	0.3208		
PRESS	178.02		Adeq Precision	7.403		

Factor	Coefficient Estimate	DF	Standard Error	95% CI Low	95% CI High	VIF
Intercept	8.50	1	0.63	7.09	9.91	
D-D	-2.25	1	0.63	-3.66	-0.84	1.00
E-E	1.22	1	0.63	-0.20	2.63	1.00
F-F	-1.19	1	0.63	-2.60	0.23	1.00
G-G	1.47	1	0.63	0.054	2.88	1.00
DF	1.19	1	0.63	-0.23	2.60	1.00
FG	1.16	1	0.63	-0.26	2.57	1.00

Final Equation in Terms of Coded Factors:

$$\text{Taste Avg} = +8.50 -2.25 * D +1.22 * E -1.19 * F +1.47 * G +1.19 * D * F +1.16 * F * G$$

The results are very similar for average taste without DCM as they were with DCM.

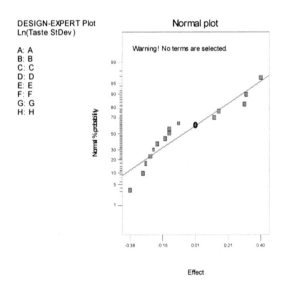

DESIGN-EXPERT Plot
Ln(Taste StDev)

A: A
B: B
C: C
D: D
E: E
F: F
G: G
H: H

The standard deviation response is much the same with or without DCM's responses. Again, there are no significant factors.

(e) Suppose that just before the start of the experiment, Harry and Judy discovered that the eight new barrels they ordered from France for use in the experiment would not arrive in time, and all 16 runs would have to be made with old barrels. If Harry and Judy just drop column C from their design, what does this do to the alias relationships? Do they need to start over and construct a new design?

The resulting design is a 2_{IV}^{7-3} with defining relations: $I = ABEF = ADEG = BDFG = ABDH = DEFH = BEGH = AFGH$.

(f) Harry and Judy know from experience that some treatment combinations are unlikely to produce good results. For example, the run with all eight variables at the high level generally results in a poorly rated wine. This was confirmed in the March 8, 1986 taste test. They want to set up a new design for their 1986 Pinot Noir using these same eight variables, but they do not want to make the run with all eight factors at the high level. What design would you suggest?

By changing the sign of any of the design generators, a design that does not include the principal fraction will be generated. This will give a design without an experimental run combination with all of the variables at the high level.

8.49. In an article in *Quality Engineering* ("An Application of Fractional Factorial Experimental Designs," 1988, Vol. 1, pp. 19-23) M.B. Kilgo describes an experiment to determine the effect of CO_2 pressure (A), CO_2 temperature (B), peanut moisture (C), CO_2 flow rate (D), and peanut particle size (E) on the total yield of oil per batch of peanuts (y). The levels she used for these factors are shown in Table P8.12.

Table P8.7

Coded Level	A Pressure (bar)	B Temp (C)	C Moisture (% by weight)	D Flow (liters/min)	E Particle Size (mm)
-1	415	25	5	40	1.28
1	550	95	15	60	4.05

She conducted the 16-run fractional factorial experiment in Table P8.8.

Table P8.8

	A	B	C	D	E	y
1	415	25	5	40	1.28	63
2	550	25	5	40	4.05	21
3	415	95	5	40	4.05	36
4	550	95	5	40	1.28	99
5	415	25	15	40	4.05	24
6	550	25	15	40	1.28	66
7	415	95	15	40	1.28	71
8	550	95	15	40	4.05	54
9	415	25	5	60	4.05	23
10	550	25	5	60	1.28	74
11	415	95	5	60	1.28	80
12	550	95	5	60	4.05	33
13	415	25	15	60	1.28	63
14	550	25	15	60	4.05	21
15	415	95	15	60	4.05	44
16	550	95	15	60	1.28	96

(a) What type of design has been used? Identify the defining relation and the alias relationships.

A 2_V^{5-1}, 16-run design, with I= -ABCDE.

A(-ABCDE)= -BCDE	A= -BCDE
B(-ABCDE)= -ACDE	B= -ACDE
C(-ABCDE)= -ABDE	C= -ABDE
D(-ABCDE)= -ABCE	D= -ABCE
E(-ABCDE)= -ABCD	E= -ABCD
AB(-ABCDE)= -CDE	AB= -CDE
AC(-ABCDE)= -BDE	AC= -BDE
AD(-ABCDE)= -BCE	AD= -BCE
AE(-ABCDE)= -BCD	AE= -BCD
BC(-ABCDE)= -ADE	BC= -ADE
BD(-ABCDE)= -ACE	BD= -ACE
BE(-ABCDE)= -ACD	BE= -ACD
CD(-ABCDE)= -ABE	CD= -ABE
CE(-ABCDE)= -ABD	CE= -ABD
DE(-ABCDE)= -ABC	DE= -ABC

(b) Estimate the factor effects and use a normal probability plot to tentatively identify the important factors.

Design-Expert Output

	Term	Effect	SumSqr	% Contribtn
Model	Intercept			
Error	A	7.5	225	2.17119
Model	B	19.75	1560.25	15.056
Error	C	1.25	6.25	0.0603107
Error	D	0	0	0
Model	E	44.5	7921	76.4354
Error	AB	5.25	110.25	1.06388
Error	AC	1.25	6.25	0.0603107
Error	AD	-4	64	0.617582
Error	AE	7	196	1.89134
Error	BC	3	36	0.34739
Error	BD	-1.75	12.25	0.118209
Error	BE	0.25	0.25	0.00241243
Error	CD	2.25	20.25	0.195407
Error	CE	-6.25	156.25	1.50777
Error	DE	3.5	49	0.472836
	Lenth's ME	11.5676		
	Lenth's SME	23.4839		

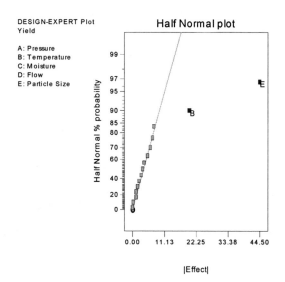

(c) Perform an appropriate statistical analysis to test the hypothesis that the factors identified in part (b) above have a significant effect on the yield of peanut oil.

Design-Expert Output

Response: Yield
ANOVA for Selected Factorial Model
Analysis of variance table [Partial sum of squares]

Source	Sum of Squares	DF	Mean Square	F Value	Prob > F	
Model	9481.25	2	4740.63	69.89	< 0.0001	significant
B	1560.25	1	1560.25	23.00	0.0003	
E	7921.00	1	7921.00	116.78	< 0.0001	
Residual	881.75	13	67.83			
Cor Total	10363.00	15				

The Model F-value of 69.89 implies the model is significant. There is only a 0.01% chance that a "Model F-Value" this large could occur due to noise.

Std. Dev.	8.24	R-Squared	0.9149	
Mean	54.25	Adj R-Squared	0.9018	
C.V.	15.18	Pred R-Squared	0.8711	
PRESS	1335.67	Adeq Precision	18.017	

Factor	Coefficient Estimate	DF	Standard Error	95% CI Low	95% CI High	VIF
Intercept	54.25	1	2.06	49.80	58.70	
B-Temperature	9.88	1	2.06	5.43	14.32	1.00
E-Particle Size	22.25	1	2.06	17.80	26.70	1.00

(d) Fit a model that could be used to predict peanut oil yield in terms of the factors that you have identified as important.

Design-Expert Output

Final Equation in Terms of Coded Factors:

Yield   =
+54.25
+9.88   * B
+22.25   * E

Final Equation in Terms of Actual Factors:

Yield   =
-5.49175
+0.28214   * Temperature
+16.06498   * Particle Size

(e) Analyze the residuals from this experiment and comment on model adequacy.

The residual plots are satisfactory. There is a slight tendency for the variability of the residuals to increase with the predicted value of *y*.

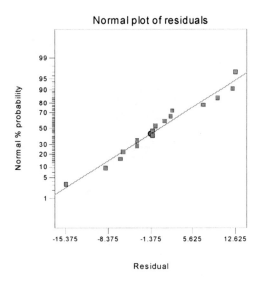

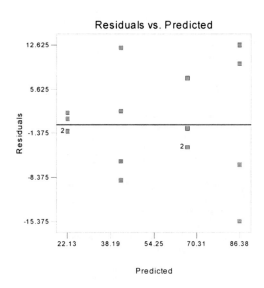

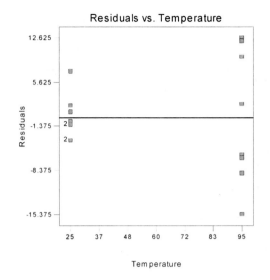

Residuals vs. Temperature

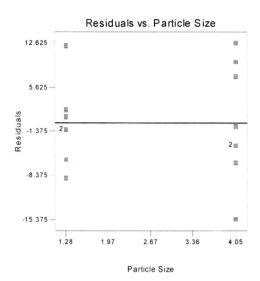

Residuals vs. Particle Size

8.51. A 16-run fractional factorial experiment in nine factors was conducted by Chrysler Motors Engineering and described in the article "Sheet Molded Compound Process Improvement," by P.I. Hsieh and D.E. Goodwin (*Fourth Symposium on Taguchi Methods*, American Supplier Institute, Dearborn, MI, 1986, pp. 13-21). The purpose was to reduce the number of defects in the finish of sheet-molded grill opening panels. The design, and the resulting number of defects, c, observed on each run, is shown in Table P8.10. This is a resolution III fraction with generators $E=BD$, $F=BCD$, $G=AC$, $H=ACD$, and $J=AB$.

Table P8.10

Run	A	B	C	D	E	F	G	H	J	c	$\sqrt{c}$	F&T's Modification
1	-	-	-	-	+	-	+	-	+	56	7.48	7.52
2	+	-	-	-	+	-	-	+	-	17	4.12	4.18
3	-	+	-	-	-	+	+	-	-	2	1.41	1.57
4	+	+	-	-	-	+	-	+	+	4	2.00	2.12
5	-	-	+	-	+	+	-	+	+	3	1.73	1.87
6	+	-	+	-	+	+	+	-	-	4	2.00	2.12
7	-	+	+	-	-	-	-	+	-	50	7.07	7.12
8	+	+	+	-	-	-	+	-	+	2	1.41	1.57
9	-	-	-	+	-	+	+	+	+	1	1.00	1.21
10	+	-	-	+	-	+	-	-	-	0	0.00	0.50
11	-	+	-	+	+	-	+	+	-	3	1.73	1.87
12	+	+	-	+	+	-	-	-	+	12	3.46	3.54
13	-	-	+	+	-	-	-	-	+	3	1.73	1.87
14	+	-	+	+	-	-	+	+	-	4	2.00	2.12
15	-	+	+	+	+	+	-	-	-	0	0.00	0.50
16	+	+	+	+	+	+	+	+	+	0	0.00	0.50

(a) Find the defining relation and the alias relationships in this design.

$I = ABJ = ACG = BDE = CEF = DGH = FHJ = ABFH = ACDH = ADEJ = AEFG = BCDF = BCGJ = BEGH = CEHJ = DFGJ = ABCEH = ABDFG = ACDFJ = ADEFH = AEGHJ = BCDHJ = BCFGH = BEFGJ = CDEGJ = ABCDEG = ABCEFJ = ABDGHJ = ACFGHJ = BDEFHJ = CDEFGH = ABCDEFGHJ$

(b) Estimate the factor effects and use a normal probability plot to tentatively identify the important factors.

The effects are shown below in the Design-Expert output. The normal probability plot of effects identifies factors A, D, F, and interactions AD, AF, BC, BG as important.

Design-Expert Output

	Term	Effect	SumSqr	% Contribtn
Model	Intercept			
Model	A	-9.375	351.562	7.75573
Model	B	-1.875	14.0625	0.310229
Model	C	-3.625	52.5625	1.15957
Model	D	-14.375	826.562	18.2346
Error	E	3.625	52.5625	1.15957
Model	F	-16.625	1105.56	24.3895
Model	G	-2.125	18.0625	0.398472
Error	H	0.375	0.5625	0.0124092
Error	J	0.125	0.0625	0.0013788
Model	AD	11.625	540.563	11.9252
Error	AE	2.125	18.0625	0.398472
Model	AF	9.875	390.063	8.60507
Error	AH	1.375	7.5625	0.166834
Model	BC	11.375	517.563	11.4178
Model	BG	-12.625	637.562	14.0651
	Lenth's ME	13.9775		
	Lenth's SME	28.3764		

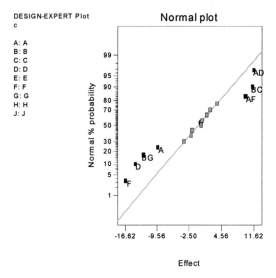

DESIGN-EXPERT Plot
c

A: A
B: B
C: C
D: D
E: E
F: F
G: G
H: H
J: J

Normal plot

(c) Fit an appropriate model using the factors identified in part (b) above.

The analysis of variance and corresponding model is shown below. Factors B, C, and G are included for hierarchal purposes.

Design-Expert Output

Response: c
   ANOVA for Selected Factorial Model
Analysis of variance table [Partial sum of squares]

Source	Sum of Squares	DF	Mean Square	F Value	Prob > F	
Model	4454.13	10	445.41	28.26	0.0009	significant
A	*351.56*	*1*	*351.56*	*22.30*	*0.0052*	
B	*14.06*	*1*	*14.06*	*0.89*	*0.3883*	
C	*52.56*	*1*	*52.56*	*3.33*	*0.1274*	
D	*826.56*	*1*	*826.56*	*52.44*	*0.0008*	
F	*1105.56*	*1*	*1105.56*	*70.14*	*0.0004*	
G	*18.06*	*1*	*18.06*	*1.15*	*0.3333*	
AD	*540.56*	*1*	*540.56*	*34.29*	*0.0021*	
AF	*390.06*	*1*	*390.06*	*24.75*	*0.0042*	
BC	*517.56*	*1*	*517.56*	*32.84*	*0.0023*	
BG	*637.56*	*1*	*637.56*	*40.45*	*0.0014*	
Residual	78.81	5	15.76			
Cor Total	4532.94	15				

The Model F-value of 28.26 implies the model is significant. There is only
a 0.09% chance that a "Model F-Value" this large could occur due to noise.

Std. Dev.	3.97	R-Squared	0.9826	
Mean	10.06	Adj R-Squared	0.9478	
C.V.	39.46	Pred R-Squared	0.8220	
PRESS	807.04	Adeq Precision	17.771	

Factor	Coefficient Estimate	DF	Standard Error	95% CI Low	95% CI High	VIF
Intercept	10.06	1	0.99	7.51	12.61	
A-A	-4.69	1	0.99	-7.24	-2.14	1.00
B-B	-0.94	1	0.99	-3.49	1.61	1.00
C-C	-1.81	1	0.99	-4.36	0.74	1.00
D-D	-7.19	1	0.99	-9.74	-4.64	1.00
F-F	-8.31	1	0.99	-10.86	-5.76	1.00
G-G	-1.06	1	0.99	-3.61	1.49	1.00
AD	5.81	1	0.99	3.26	8.36	1.00
AF	4.94	1	0.99	2.39	7.49	1.00
BC	5.69	1	0.99	3.14	8.24	1.00
BG	-6.31	1	0.99	-8.86	-3.76	1.00

Final Equation in Terms of Coded Factors:

$$
\begin{aligned}
c = \ &+10.06 \\
&-4.69 \quad * A \\
&-0.94 \quad * B \\
&-1.81 \quad * C \\
&-7.19 \quad * D \\
&-8.31 \quad * F \\
&-1.06 \quad * G \\
&+5.81 \quad * A * D \\
&+4.94 \quad * A * F \\
&+5.69 \quad * B * C \\
&-6.31 \quad * B * G
\end{aligned}
$$

Final Equation in Terms of Actual Factors:

	c	=
	+10.06250	
	-4.68750	* A
	-0.93750	* B
	-1.81250	* C
	-7.18750	* D
	-8.31250	* F
	-1.06250	* G
	+5.81250	* A * D
	+4.93750	* A * F
	+5.68750	* B * C
	-6.31250	* B * G

(d) Plot the residuals from this model versus the predicted number of defects. Also, prepare a normal probability plot of the residuals. Comment on the adequacy of these plots.

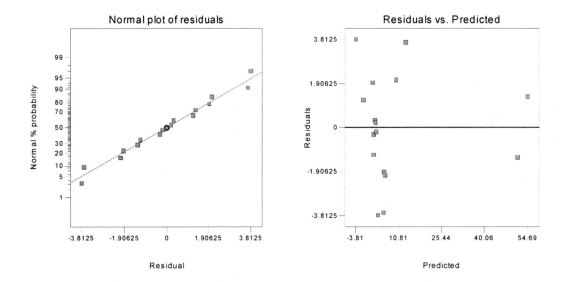

There is a significant problem with inequality of variance. This is likely caused by the response variable being a count. A transformation may be appropriate.

(e) In part (d) you should have noticed an indication that the variance of the response is not constant (considering that the response is a count, you should have expected this). The previous table also shows a transformation on c, the square root, that is a widely used variance stabilizing transformation for count data (refer to the discussion of variance stabilizing transformations in Chapter 3). Repeat parts (a) through (d) using the transformed response and comment on your results. Specifically, are the residual plots improved?

Design-Expert Output

	Term	Effect	SumSqr	% Contribtn
Model	Intercept			
Error	A	-0.895	3.2041	4.2936
Model	B	-0.3725	0.555025	0.743752
Error	C	-0.6575	1.72922	2.31722
Model	D	-2.1625	18.7056	25.0662
Error	E	0.4875	0.950625	1.27387
Model	F	-2.6075	27.1962	36.4439
Model	G	-0.385	0.5929	0.794506
Error	H	0.27	0.2916	0.390754
Error	J	0.06	0.0144	0.0192965
Error	AD	1.145	5.2441	7.02727
Error	AE	0.555	1.2321	1.65106
Error	AF	0.86	2.9584	3.96436
Error	AH	0.0425	0.007225	0.00968175
Error	BC	0.6275	1.57502	2.11059
Model	BG	-1.61	10.3684	13.894
	Lenth's ME	2.27978		
	Lenth's SME	4.62829		

The analysis of the data with the square root transformation identifies only D, F, the BG interaction as being significant. The original analysis identified factor A and several two factor interactions as being significant.

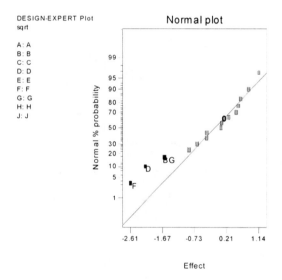

Design-Expert Output

Response: sqrt
ANOVA for Selected Factorial Model
Analysis of variance table [Partial sum of squares]

Source	Sum of Squares	DF	Mean Square	F Value	Prob > F	
Model	57.42	5	11.48	6.67	0.0056	significant
B	0.56	1	0.56	0.32	0.5826	
D	18.71	1	18.71	10.87	0.0081	
F	27.20	1	27.20	15.81	0.0026	
G	0.59	1	0.59	0.34	0.5702	
BG	10.37	1	10.37	6.03	0.0340	
Residual	17.21	10	1.72			
Cor Total	74.62	15				

The Model F-value of 6.67 implies the model is significant. There is only a 0.56% chance that a "Model F-Value" this large could occur due to noise.

Std. Dev.	1.31	R-Squared	0.7694
Mean	2.32	Adj R-Squared	0.6541
C.V.	56.51	Pred R-Squared	0.4097
PRESS	44.05	Adeq Precision	8.422

Factor	Coefficient Estimate	DF	Standard Error	95% CI Low	95% CI High	VIF
Intercept	2.32	1	0.33	1.59	3.05	
B-B	-0.19	1	0.33	-0.92	0.54	1.00
D-D	-1.08	1	0.33	-1.81	-0.35	1.00
F-F	-1.30	1	0.33	-2.03	-0.57	1.00
G-G	-0.19	1	0.33	-0.92	0.54	1.00
BG	-0.80	1	0.33	-1.54	-0.074	1.00

Final Equation in Terms of Coded Factors:

$$
\begin{aligned}
\text{sqrt} = \quad & +2.32 \\
& -0.19 \quad * B \\
& -1.08 \quad * D \\
& -1.30 \quad * F \\
& -0.19 \quad * G \\
& -0.80 \quad * B * G
\end{aligned}
$$

Final Equation in Terms of Actual Factors:

$$
\begin{aligned}
\text{sqrt} = \quad & +2.32125 \\
& -0.18625 \quad * B \\
& -1.08125 \quad * D \\
& -1.30375 \quad * F \\
& -0.19250 \quad * G \\
& -0.80500 \quad * B * G
\end{aligned}
$$

The residual plots are acceptable, although there appears to be a slight "u" shape to the residuals versus predicted plot.

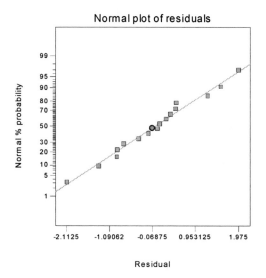

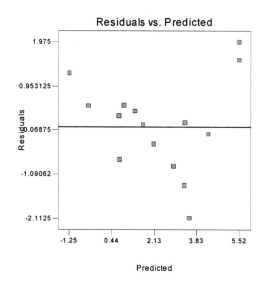

(f) There is a modification to the square root transformation proposed by Freeman and Tukey ("Transformations Related to the Angular and the Square Root," *Annals of Mathematical Statistics*, Vol. 21, 1950, pp. 607-611) that improves its performance. F&T's modification to the square root transformation is:

$$\frac{1}{2}\left[\sqrt{c} + \sqrt{c+1}\right]$$

Rework parts (a) through (d) using this transformation and comment on the results. (For an interesting discussion and analysis of this experiment, refer to "Analysis of Factorial Experiments with Defects or Defectives as the Response," by S. Bisgaard and H.T. Fuller, *Quality Engineering*, Vol. 7, 1994-5, pp. 429-443.)

Design-Expert Output

	Term	Effect	SumSqr	% Contribtn
Model	Intercept			
Error	A	-0.86	2.9584	4.38512
Model	B	-0.325	0.4225	0.626255
Error	C	-0.605	1.4641	2.17018
Model	D	-1.995	15.9201	23.5977
Error	E	0.5025	1.01002	1.49712
Model	F	-2.425	23.5225	34.8664
Model	G	-0.4025	0.648025	0.960541
Error	H	0.225	0.2025	0.300158
Error	J	0.0275	0.003025	0.00448383
Error	AD	1.1625	5.40562	8.01254
Error	AE	0.505	1.0201	1.51205
Error	AF	0.8825	3.11523	4.61757
Error	AH	0.0725	0.021025	0.0311645
Error	BC	0.7525	2.26503	3.35735
Model	BG	-1.54	9.4864	14.0613
	Lenth's ME	2.14001		
	Lenth's SME	4.34453		

As with the square root transformation, factors *D*, *F*, and the *BG* interaction remain significant.

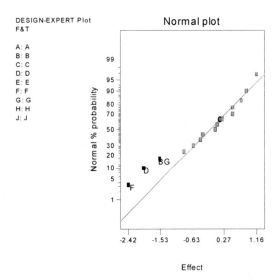

Design-Expert Output

Response: F&T
ANOVA for Selected Factorial Model
Analysis of variance table [Partial sum of squares]

Source	Sum of Squares	DF	Mean Square	F Value	Prob > F	
Model	50.00	5	10.00	5.73	0.0095	significant
B	0.42	1	0.42	0.24	0.6334	
D	15.92	1	15.92	9.12	0.0129	
F	23.52	1	23.52	13.47	0.0043	
G	0.65	1	0.65	0.37	0.5560	
BG	9.49	1	9.49	5.43	0.0420	
Residual	17.47	10	1.75			
Cor Total	67.46	15				

The Model F-value of 5.73 implies the model is significant. There is only
a 0.95% chance that a "Model F-Value" this large could occur due to noise.

Std. Dev.	1.32	R-Squared	0.7411	
Mean	2.51	Adj R-Squared	0.6117	
C.V.	52.63	Pred R-Squared	0.3373	
PRESS	44.71	Adeq Precision	7.862	

Factor	Coefficient Estimate	DF	Standard Error	95% CI Low	95% CI High	VIF
Intercept	2.51	1	0.33	1.78	3.25	
B-B	-0.16	1	0.33	-0.90	0.57	1.00
D-D	-1.00	1	0.33	-1.73	-0.26	1.00
F-F	-1.21	1	0.33	-1.95	-0.48	1.00
G-G	-0.20	1	0.33	-0.94	0.53	1.00
BG	-0.77	1	0.33	-1.51	-0.034	1.00

Final Equation in Terms of Coded Factors:

$$F\&T =$$
+2.51
-0.16 * B
-1.00 * D
-1.21 * F
-0.20 * G
-0.77 * B * G

Final Equation in Terms of Actual Factors:

$$F\&T =$$
+2.51125
-0.16250 * B
-0.99750 * D
-1.21250 * F
-0.20125 * G
-0.77000 * B * G

The following interaction plots appear as they did with the square root transformation; a slight "u" shape is observed in the residuals versus predicted plot.

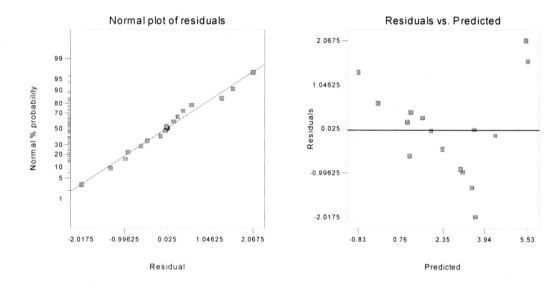

8.52. An experiment is run in a semiconductor factory to investigate the effect of six factors on transistor gain. The design selected is the 2_{IV}^{6-2} shown in Table P8.15.

Table P8.11

Standard Order	Run Order	A	B	C	D	E	F	Gain
1	2	-	-	-	-	-	-	1455
2	8	+	-	-	-	+	-	1511
3	5	-	+	-	-	+	+	1487
4	9	+	+	-	-	-	+	1596
5	3	-	-	+	-	+	+	1430
6	14	+	-	+	-	-	+	1481
7	11	-	+	+	-	-	-	1458
8	10	+	+	+	-	+	-	1549
9	15	-	-	-	+	-	+	1454
10	13	+	-	-	+	+	+	1517
11	1	-	+	-	+	+	-	1487
12	6	+	+	-	+	-	-	1596
13	12	-	-	+	+	+	-	1446
14	4	+	-	+	+	-	-	1473
15	7	-	+	+	+	-	+	1461
16	16	+	+	+	+	+	+	1563

(a) Use a normal plot of the effects to identify the significant factors.

Design-Expert Output

	Term	Effect	SumSqr	% Contribtn
Model	Intercept			
Model	A	76	23104	55.2714
Model	B	53.75	11556.2	27.6459
Model	C	-30.25	3660.25	8.75637
Error	D	3.75	56.25	0.134566
Error	E	2	16	0.0382766
Error	F	1.75	12.25	0.0293055
Model	AB	26.75	2862.25	6.84732
Model	AC	-8.25	272.25	0.6513
Error	AD	-0.75	2.25	0.00538265
Error	AE	-3.5	49	0.117222
Error	AF	5.25	110.25	0.26375
Error	BD	0.5	1	0.00239229
Error	BF	2.5	25	0.0598072
Error	ABD	3.5	49	0.117222
Error	ABF	-2.5	25	0.0598072
	Lenth's ME	9.63968		
	Lenth's SME	19.57		

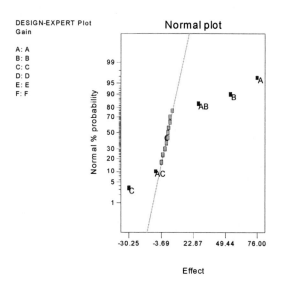

(b) Conduct appropriate statistical tests for the model identified in part (a).

Design-Expert Output

Response: Gain
ANOVA for Selected Factorial Model
Analysis of variance table [Partial sum of squares]

Source	Sum of Squares	DF	Mean Square	F Value	Prob > F	
Model	41455.00	5	8291.00	239.62	< 0.0001	significant
A	23104.00	1	23104.00	667.75	< 0.0001	
B	11556.25	1	11556.25	334.00	< 0.0001	
C	3660.25	1	3660.25	105.79	< 0.0001	
AB	2862.25	1	2862.25	82.72	< 0.0001	
AC	272.25	1	272.25	7.87	0.0186	
Residual	346.00	10	34.60			
Cor Total	41801.00	15				

The Model F-value of 239.62 implies the model is significant. There is only
a 0.01% chance that a "Model F-Value" this large could occur due to noise.

Std. Dev.	5.88	R-Squared	0.9917	
Mean	1497.75	Adj R-Squared	0.9876	
C.V.	0.39	Pred R-Squared	0.9788	
PRESS	885.76	Adeq Precision	44.419	

Factor	Coefficient Estimate	DF	Standard Error	95% CI Low	95% CI High	VIF
Intercept	1497.75	1	1.47	1494.47	1501.03	
A-A	38.00	1	1.47	34.72	41.28	1.00
B-B	26.87	1	1.47	23.60	30.15	1.00
C-C	-15.13	1	1.47	-18.40	-11.85	1.00
AB	13.38	1	1.47	10.10	16.65	1.00
AC	-4.12	1	1.47	-7.40	-0.85	1.00

Final Equation in Terms of Coded Factors:

Gain   =
+1497.75
+38.00   * A
+26.87   * B
-15.13   * C
+13.38   * A * B
-4.12   * A * C

Final Equation in Terms of Actual Factors:

Gain   =
+1497.75000
+38.00000   * A
+26.87500   * B
-15.12500   * C
+13.37500   * A * B
-4.12500   * A * C

(c) Analyze the residuals and comment on your findings.

The residual plots are acceptable. The normality and equality of variance assumptions are verified. There do not appear to be any trends or interruptions in the residuals versus run order plot.

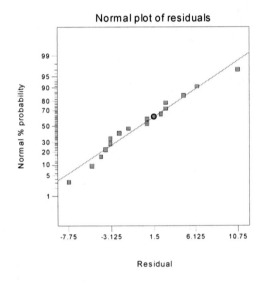

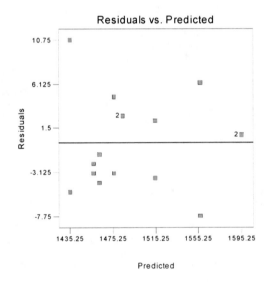

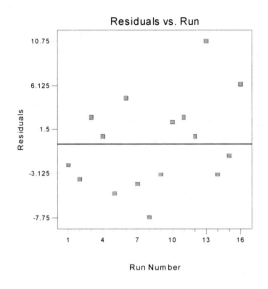

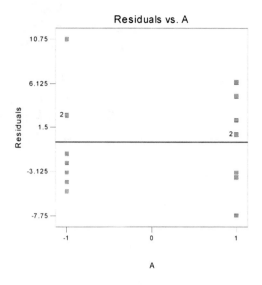

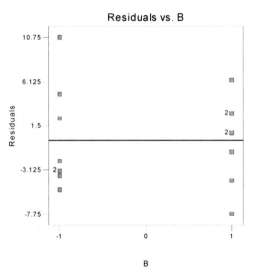

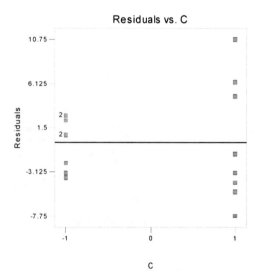

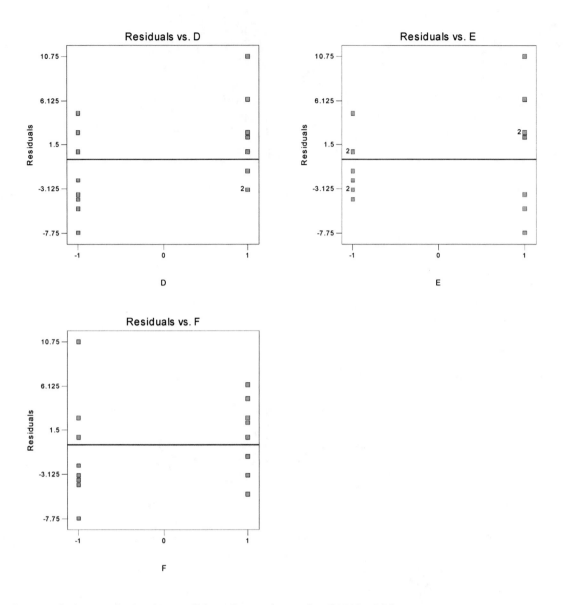

(d) Can you find a set of operating conditions that produce gain of 1500 ± 25 ?

Yes, see the graphs below.

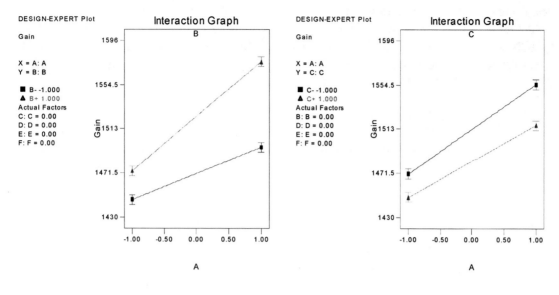

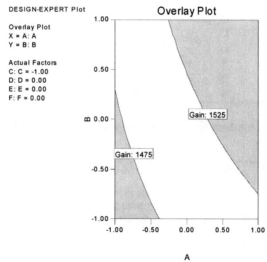

8.53. Heat treating is often used to carbonize metal parts, such as gears. The thickness of the carbonized layer is a critical output variable from this process, and it is usually measured by performing a carbon analysis on the gear pitch (top of the gear tooth). Six factors were studied on a 2_{IV}^{6-2} design: A = furnace temperature, B = cycle time, C = carbon concentration, D = duration of the carbonizing cycle, E = carbon concentration of the diffuse cycle, and F = duration of the diffuse cycle. The experiment is shown in Table P8.16.

Table P8.12

Standard Order	Run Order	A	B	C	D	E	F	Pitch
1	5	-	-	-	-	-	-	74
2	7	+	-	-	-	+	-	190
3	8	-	+	-	-	+	+	133
4	2	+	+	-	-	-	+	127
5	10	-	-	+	-	+	+	115
6	12	+	-	+	-	-	+	101
7	16	-	+	+	-	-	-	54
8	1	+	+	+	-	+	-	144
9	6	-	-	-	+	-	+	121
10	9	+	-	-	+	+	+	188
11	14	-	+	-	+	+	-	135
12	13	+	+	-	+	-	-	170
13	11	-	-	+	+	+	-	126
14	3	+	-	+	+	-	-	175
15	15	-	+	+	+	-	+	126
16	4	+	+	+	+	+	+	193

(a) Estimate the factor effects and plot them on a normal probability plot. Select a tentative model.

Design-Expert Output

	Term	Effect	SumSqr	% Contribtn
Model	Intercept			
Model	A	50.5	10201	41.8777
Error	B	-1	4	0.016421
Model	C	-13	676	2.77515
Model	D	37	5476	22.4804
Model	E	34.5	4761	19.5451
Error	F	4.5	81	0.332526
Error	AB	-4	64	0.262737
Error	AC	-2.5	25	0.102631
Error	AD	4	64	0.262737
Error	AE	1	4	0.016421
Error	BD	4.5	81	0.332526
Model	CD	14.5	841	3.45252
Model	DE	-22	1936	7.94778
Error	ABD	0.5	1	0.00410526
Error	ABF	6	144	0.591157
	Lenth's ME	15.4235		
	Lenth's SME	31.3119		

Factors A, C, D, E and the two factor interactions CD and DE appear to be significant. The CD and DE interactions are aliased with BF and AF interactions respectively. Because factors B and F are not significant, CD and DE were included in the model. The model can be found in the Design-Expert Output below.

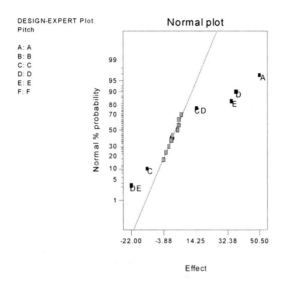

DESIGN-EXPERT Plot
Pitch

A: A
B: B
C: C
D: D
E: E
F: F

(b) Perform appropriate statistical tests on the model.

Design-Expert Output

Response: Pitch
ANOVA for Selected Factorial Model
Analysis of variance table [Partial sum of squares]

Source	Sum of Squares	DF	Mean Square	F Value	Prob > F	
Model	23891.00	6	3981.83	76.57	< 0.0001	significant
A	*10201.00*	*1*	*10201.00*	*196.17*	*< 0.0001*	
C	*676.00*	*1*	*676.00*	*13.00*	*0.0057*	
D	*5476.00*	*1*	*5476.00*	*105.31*	*< 0.0001*	
E	*4761.00*	*1*	*4761.00*	*91.56*	*< 0.0001*	
CD	*841.00*	*1*	*841.00*	*16.17*	*0.0030*	
DE	*1936.00*	*1*	*1936.00*	*37.23*	*0.0002*	
Residual	468.00	9	52.00			
Cor Total	24359.00	15				

The Model F-value of 76.57 implies the model is significant. There is only a 0.01% chance that a "Model F-Value" this large could occur due to noise.

Std. Dev.	7.21	R-Squared	0.9808	
Mean	135.75	Adj R-Squared	0.9680	
C.V.	5.31	Pred R-Squared	0.9393	
PRESS	1479.11	Adeq Precision	28.618	

Factor	Coefficient Estimate	DF	Standard Error	95% CI Low	95% CI High	VIF
Intercept	135.75	1	1.80	131.67	139.83	
A-A	25.25	1	1.80	21.17	29.33	1.00
C-C	-6.50	1	1.80	-10.58	-2.42	1.00
D-D	18.50	1	1.80	14.42	22.58	1.00
E-E	17.25	1	1.80	13.17	21.33	1.00
CD	7.25	1	1.80	3.17	11.33	1.00
DE	-11.00	1	1.80	-15.08	-6.92	1.00

Final Equation in Terms of Coded Factors:

Pitch   =
+135.75
+25.25   * A
-6.50   * C
+18.50   * D
+17.25   * E
+7.25   * C * D
-11.00   * D * E

Final Equation in Terms of Actual Factors:

Pitch   =
+135.75000
+25.25000   * A
-6.50000   * C
+18.50000   * D
+17.25000   * E
+7.25000   * C * D
-11.00000   * D * E

(c) Analyze the residuals and comment on model adequacy.

The residual plots are acceptable. The normality and equality of variance assumptions are verified. There do not appear to be any trends or interruptions in the residuals versus run order plot. The plots of the residuals versus factors C and E identify reduced variation at the lower level of both variables, while the plot of residuals versus factor F identifies reduced variation at the upper level. Because C and E are significant factors in the model, this might not affect the decision on the optimum solution for the process. However, factor F is not included in the model and may be set at the upper level to reduce variation.

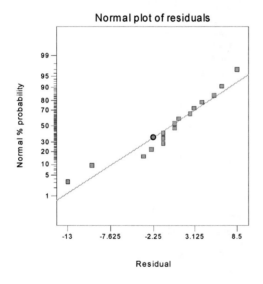

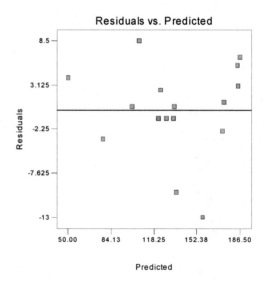

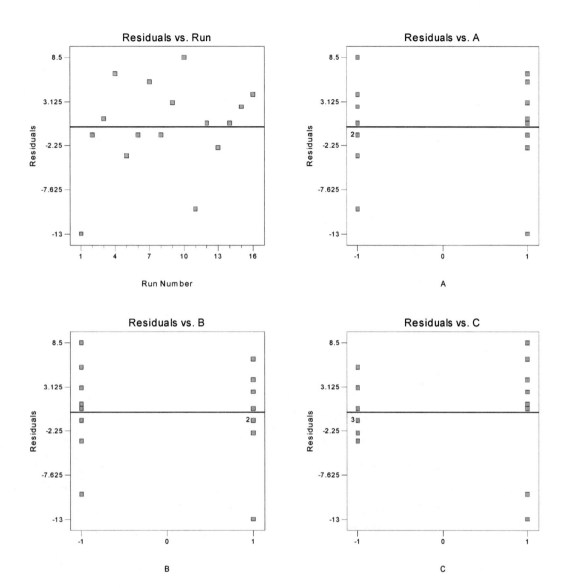

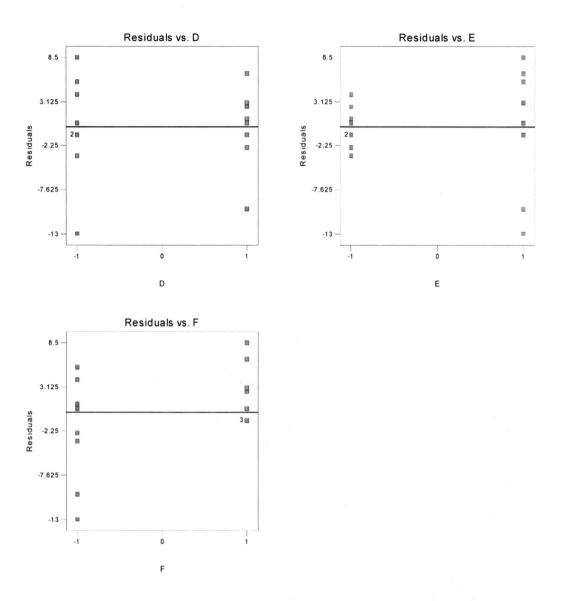

(d) Interpret the results of this experiment. Assume that a layer thickness of between 140 and 160 is desirable.

The following graphs identify a region that is acceptable between 140 and 160.

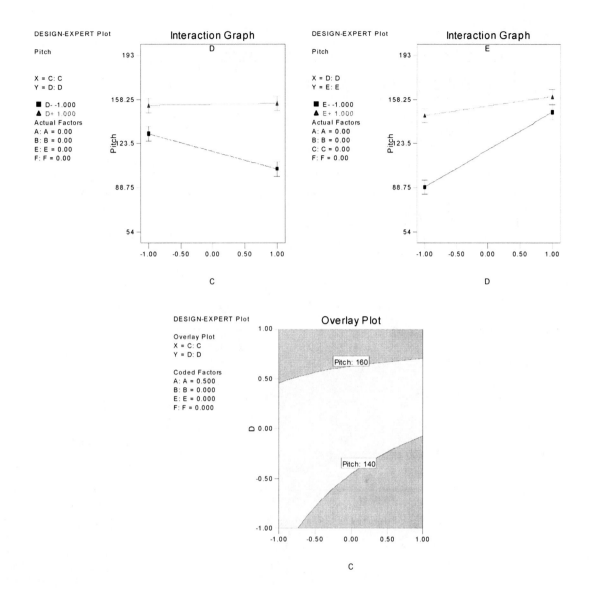

8.54. An article by L.B. Hare ("In the Soup: A Case Study to Identify Contributors to Filling Variability," *Journal of Quality Technology*, Vol. 20, pp. 36-43) describes a factorial experiment used to study the filling variability of dry soup mix packages. The factors are A = number of mixing ports through which the vegetable oil was added (1, 2), B = temperature surrounding the mixer (cooled, ambient), C = mixing time (60, 80 sec), D = batch weight (1500, 2000 lb), and E = number of days between mixing and packaging (1,7). Between 125 and 150 packages of soup were sampled over an eight hour period for each run in the design and the standard deviation of package weight was used as the response variable. The design and resulting data are shown in Table P8.17.

Table P8.13

Std Order	A - Mixer Ports	B - Temp	C - Time	D - Batch Weight	E - Delay	y - Std Dev
1	-1	-1	-1	-1	-1	1.13
2	1	-1	-1	-1	1	1.25
3	-1	1	-1	-1	1	0.97
4	1	1	-1	-1	-1	1.70
5	-1	-1	1	-1	1	1.47
6	1	-1	1	-1	-1	1.28
7	-1	1	1	-1	-1	1.18
8	1	1	1	-1	1	0.98
9	-1	-1	-1	1	1	0.78
10	1	-1	-1	1	-1	1.36
11	-1	1	-1	1	-1	1.85
12	1	1	-1	1	1	0.62
13	-1	-1	1	1	-1	1.09
14	1	-1	1	1	1	1.10
15	-1	1	1	1	1	0.76
16	1	1	1	1	-1	2.10

(a) What is the generator for this design?

The design generator is $I = -ABCDE$.

(b) What is the resolution of this design?

This design is Resolution V.

(c) Estimate the factor effects. Which effects are large?

Design-Expert Output

	Term	Effect	SumSqr	% Contribtn
Require	Intercept			
Error	A	0.145	0.0841	3.48388
Model	B	0.0875	0.030625	1.26865
Error	C	0.0375	0.005625	0.233018
Model	D	-0.0375	0.005625	0.233018
Model	E	-0.47	0.8836	36.6035
Error	AB	0.015	0.0009	0.0372829
Error	AC	0.095	0.0361	1.49546
Error	AD	0.03	0.0036	0.149132
Error	AE	-0.1525	0.093025	3.8536
Error	BC	-0.0675	0.018225	0.754979
Error	BD	0.1625	0.105625	4.37556
Model	BE	-0.405	0.6561	27.1792
Error	CD	0.0725	0.021025	0.87097
Error	CE	0.135	0.0729	3.01992
Model	DE	-0.315	0.3969	16.4418

Lenth's ME   0.337389
Lenth's SME 0.684948

Factor *E* and the two factor interactions *BE* and *DE* appear to be significant. Factors *B* and *D* are included to satisfy model hierarchy. The analysis of variance and model can be found in the Design-Expert Output below.

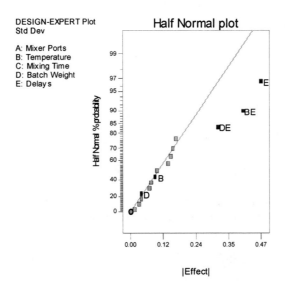

Design-Expert Output

Response: Std Dev
    ANOVA for Selected Factorial Model
Analysis of variance table [Partial sum of squares]

Source	Sum of Squares	DF	Mean Square	F Value	Prob > F	
Model	1.97	5	0.39	8.94	0.0019	significant
B	0.031	1	0.031	0.69	0.4242	
D	5.625E-003	1	5.625E-003	0.13	0.7284	
E	0.88	1	0.88	20.03	0.0012	
BE	0.66	1	0.66	14.87	0.0032	
DE	0.40	1	0.40	9.00	0.0134	
Residual	0.44	10	0.044			
Cor Total	2.41	15				

The Model F-value of 8.94 implies the model is significant. There is only
a 0.19% chance that a "Model F-Value" this large could occur due to noise.

Std. Dev.	0.21	R-Squared	0.8173
Mean	1.23	Adj R-Squared	0.7259
C.V.	17.13	Pred R-Squared	0.5322
PRESS	1.13	Adeq Precision	9.252

Final Equation in Terms of Coded Factors:

$$
\begin{aligned}
\text{Std Dev} = \\
+1.23 \\
+0.044 \quad * B \\
-0.019 \quad * D \\
-0.24 \quad * E \\
-0.20 \quad * B * E \\
-0.16 \quad * D * E
\end{aligned}
$$

Final Equation in Terms of Actual Factors:

Temperature	Cool
Std Dev	=
-0.11292	
+7.65000E-004	* Batch Weight
+0.35667	* Delays
-2.10000E-004	* Batch Weight * Delays
Temperature	Ambient
Std Dev	=
+0.51458	
+7.65000E-004	* Batch Weight
+0.22167	* Delays
-2.10000E-004	* Batch Weight * Delays

(d) Does a residual analysis indicate any problems with the underlying assumptions?

Often a transformation such as the natural log is required for the standard deviation response; however, the following residuals appear to be acceptable without the transformation.

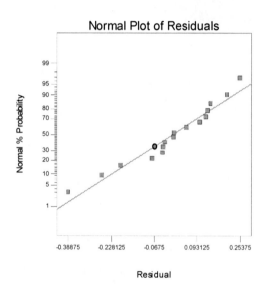

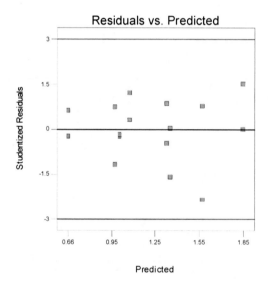

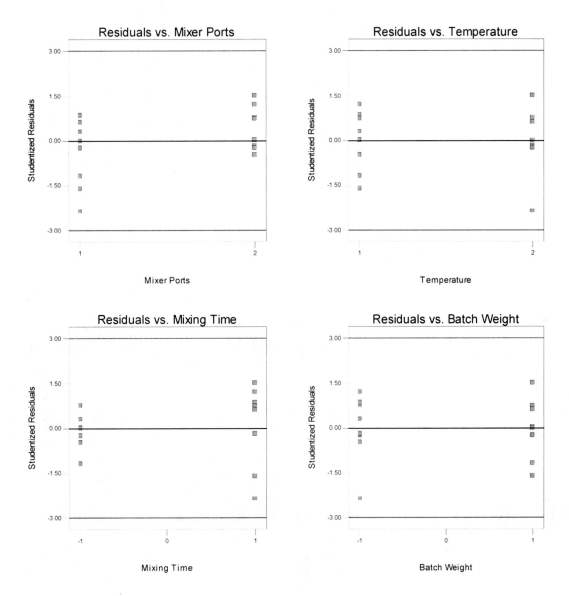

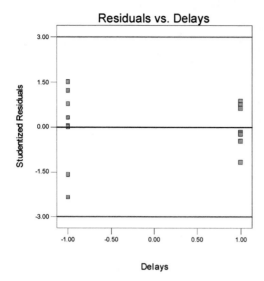

(e) Draw conclusions about this filling process.

From the interaction plots below, the lowest standard deviation can be achieved with the Temperature at ambient, Batch Weight at 2000 lbs, and a Delay of 7 days.

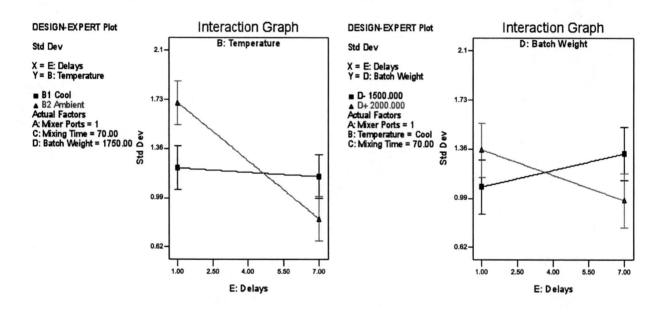

8.56. Consider the 2_{IV}^{7-3} design. Suppose that a fold over of this design is run by changing the signs in column A. Determine the alias relationship in the combined design.

MINITAB Output

```
Fractional Factorial Design

Factors:   7   Base Design:          7, 16   Resolution:   IV
Runs:     32   Replicates:              1   Fraction:    1/4
Blocks:    1   Center pts (total):      0

Design Generators (before folding): E = ABC, F = BCD, G = ACD

Folded on Factors: A

Alias Structure

I + BCDF + BDEG + CEFG

A + ABCDF + ABDEG + ACEFG
B + CDF + DEG + BCEFG
C + BDF + EFG + BCDEG
D + BCF + BEG + CDEFG
E + BDG + CFG + BCDEF
F + BCD + CEG + BDEFG
G + BDE + CEF + BCDFG
AB + ACDF + ADEG + ABCEFG
AC + ABDF + AEFG + ABCDEG
AD + ABCF + ABEG + ACDEFG
AE + ABDG + ACFG + ABCDEF
AF + ABCD + ACEG + ABDEFG
AG + ABDE + ACEF + ABCDFG
BC + DF + BEFG + CDEG
BD + CF + EG + BCDEFG
BE + DG + BCFG + CDEF
BF + CD + BCEG + DEFG
BG + DE + BCEF + CDFG
CE + FG + BCDG + BDEF
CG + EF + BCDE + BDFG
ABC + ADF + ABEFG + ACDEG
ABD + ACF + AEG + ABCDEFG
ABE + ADG + ABCFG + ACDEF
ABF + ACD + ABCEG + ADEFG
ABG + ADE + ABCEF + ACDFG
ACE + AFG + ABCDG + ABDEF
ACG + AEF + ABCDE + ABDFG
BCE + BFG + CDG + DEF
BCG + BEF + CDE + DFG
ABCE + ABFG + ACDG + ADEF
ABCG + ABEF + ACDE + ADFG
```

8.57. Reconsider the 2_{IV}^{7-3} design in Problem 8.56.

(a) Suppose that a fold over of this design is run by changing the signs in column B. Determine the alias relationship in the combined design.

MINITAB Output

Fractional Factorial Design

Factors:	7	Base Design:	7, 16	Resolution:	IV
Runs:	32	Replicates:	1	Fraction:	1/4
Blocks:	1	Center pts (total):	0		

Design Generators (before folding): E = ABC, F = BCD, G = ACD

Folded on Factors: B

Alias Structure

I + ACDG + ADEF + CEFG

```
A + CDG + DEF + ACEFG
B + ABCDG + ABDEF + BCEFG
C + ADG + EFG + ACDEF
D + ACG + AEF + CDEFG
E + ADF + CFG + ACDEG
F + ADE + CEG + ACDFG
G + ACD + CEF + ADEFG
AB + BCDG + BDEF + ABCEFG
AC + DG + AEFG + CDEF
AD + CG + EF + ACDEFG
AE + DF + ACFG + CDEG
AF + DE + ACEG + CDFG
AG + CD + ACEF + DEFG
BC + ABDG + BEFG + ABCDEF
BD + ABCG + ABEF + BCDEFG
BE + ABDF + BCFG + ABCDEG
BF + ABDE + BCEG + ABCDFG
BG + ABCD + BCEF + ABDEFG
CE + FG + ACDF + ADEG
CF + EG + ACDE + ADFG
ABC + BDG + ABEFG + BCDEF
ABD + BCG + BEF + ABCDEFG
ABE + BDF + ABCFG + BCDEG
ABF + BDE + ABCEG + BCDFG
ABG + BCD + ABCEF + BDEFG
ACE + AFG + CDF + DEG
ACF + AEG + CDE + DFG
BCE + BFG + ABCDF + ABDEG
BCF + BEG + ABCDE + ABDFG
ABCE + ABFG + BCDF + BDEG
ABCF + ABEG + BCDE + BDFG
```

(b) Compare the aliases from this combined design to those from the combined design from Problem 8.35. What differences resulted by changing the signs in a different column?

Both combined designs are still resolution *IV*, there are some two-factor interactions aliased with other two-factor interactions. In the combined design folded on *A*, all two-factor interactions with *A* are now aliased with four-factor or higher interactions. The sparsity of effects principle would tell us that the higher order interactions are highly unlikely to occur. In the combined design folded on *B*, all two-factor interactions with *B* are now aliased with four-factor or higher interactions.

8.58. Consider the 2^{7-3}_{IV} design.

(a) Suppose that a partial fold over of this design is run using column A (+ signs only). Determine the alias relationship in the combined design.

By choosing a fold over design in Design-Expert, sorting on column A and deleting the rows with a minus sign for A in the second block, the alias structures are identified below.

Design-Expert Output

```
Factorial Effects Aliases
[Est Terms]      Aliased Terms
[Intercept] = Intercept + ABCE + ABFG + ACDG + ADEF + BCDF + BDEG + CEFG
[A] = A - ABCE - ABFG - ACDG - ADEF + ABCDF + ABDEG + ACEFG
[B] = B + ACE + AFG + CDF + DEG + ABCDG + ABDEF + BCEFG
[C] = C + ABE + ADG + BDF + EFG + ABCFG + ACDEF + BCDEG
[D] = D + ACG + AEF + BCF + BEG + ABCDE + ABDFG + CDEFG
[E] = E + ABC + ADF + BDG + CFG + ABEFG + ACDEG + BCDEF
[F] = F + ABG + ADE + BCD + CEG + ABCEF + ACDFG + BDEFG
[G] = G + ABF + ACD + BDE + CEF + ABCEG + ADEFG + BCDFG
[AB] = AB - ACE - AFG + ACDF + ADEG - ABCDG - ABDEF + ABCEFG
[AC] = AC - ABE - ADG + ABDF + AEFG - ABCFG - ACDEF + ABCDEG
[AD] = AD - ACG - AEF + ABCF + ABEG - ABCDE - ABDFG + ACDEFG
[AE] = AE - ABC - ADF + ABDG + ACFG - ABEFG - ACDEG + ABCDEF
[AF] = AF - ABG - ADE + ABCD + ACEG - ABCEF - ACDFG + ABDEFG
[AG] = AG - ABF - ACD + ABDE + ACEF - ABCEG - ADEFG + ABCDFG
[BC] = BC + DF + ABC + ADF + BEFG + CDEG + ABEFG + ACDEG
[BD] = BD + CF + EG + ABCG + ABEF + ACDE + ADFG + BCDEFG
[BE] = BE + DG + ABE + ADG + BCFG + CDEF + ABCFG + ACDEF
[BF] = BF + CD + ABF + ACD + BCEG + DEFG + ABCEG + ADEFG
[BG] = BG + DE + ABG + ADE + BCEF + CDFG + ABCEF + ACDFG
[CE] = CE + FG + ACE + AFG + BCDG + BDEF + ABCDG + ABDEF
[CG] = CG + EF + ACG + AEF + BCDE + BDFG + ABCDE + ABDFG
[ABD] = ABD + ACF + AEG - ABCG - ABEF - ACDE - ADFG + ABCDEFG
[BCE] = BCE + BFG + CDG + DEF + ABCE + ABFG + ACDG + ADEF
[BCG] = BCG + BEF + CDE + DFG + ABCG + ABEF + ACDE + ADFG

Factorial Effects Defining Contrast
I = BCDF = BDEG = CEFG
```

(b) Rework part (a) using the negative signs to define the partial fold over. Does it make any difference which signs are used to define the partial fold over?

Both partial fold over designs produce the same alias relationships as shown below.

Design-Expert Output

Factorial Effects Aliases
[Est Terms]   **Aliased Terms**
[Intercept] = Intercept + ABCE + ABFG + ACDG + ADEF + BCDF + BDEG + CEFG
[A] = A - ABCE - ABFG - ACDG - ADEF + ABCDF + ABDEG + ACEFG
[B] = B + ACE + AFG + CDF + DEG + ABCDG + ABDEF + BCEFG
[C] = C + ABE + ADG + BDF + EFG + ABCFG + ACDEF + BCDEG
[D] = D + ACG + AEF + BCF + BEG + ABCDE + ABDFG + CDEFG
[E] = E + ABC + ADF + BDG + CFG + ABEFG + ACDEG + BCDEF
[F] = F + ABG + ADE + BCD + CEG + ABCEF + ACDFG + BDEFG
[G] = G + ABF + ACD + BDE + CEF + ABCEG + ADEFG + BCDFG
[AB] = AB - ACE - AFG + ACDF + ADEG - ABCDG - ABDEF + ABCEFG
[AC] = AC - ABE - ADG + ABDF + AEFG - ABCFG - ACDEF + ABCDEG
[AD] = AD - ACG - AEF + ABCF + ABEG - ABCDE - ABDFG + ACDEFG
[AE] = AE - ABC - ADF + ABDG + ACFG - ABEFG - ACDEG + ABCDEF
[AF] = AF - ABG - ADE + ABCD + ACEG - ABCEF - ACDFG + ABDEFG
[AG] = AG - ABF - ACD + ABDE + ACEF - ABCEG - ADEFG + ABCDFG
[BC] = BC + DF + ABC + ADF + BEFG + CDEG + ABEFG + ACDEG
[BD] = BD + CF + EG + ABCG + ABEF + ACDE + ADFG + BCDEFG
[BE] = BE + DG + ABE + ADG + BCFG + CDEF + ABCFG + ACDEF
[BF] = BF + CD + ABF + ACD + BCEG + DEFG + ABCEG + ADEFG
[BG] = BG + DE + ABG + ADE + BCEF + CDFG + ABCEF + ACDFG
[CE] = CE + FG + ACE + AFG + BCDG + BDEF + ABCDG + ABDEF
[CG] = CG + EF + ACG + AEF + BCDE + BDFG + ABCDE + ABDFG
[ABD] = ABD + ACF + AEG - ABCG - ABEF - ACDE - ADFG + ABCDEFG
[BCE] = BCE + BFG + CDG + DEF + ABCE + ABFG + ACDG + ADEF
[BCG] = BCG + BEF + CDE + DFG + ABCG + ABEF + ACDE + ADFG

Factorial Effects Defining Contrast
I = BCDF = BDEG = CEFG

CHAPTER **9**

Three-Level and Mixed-Level Factorial and Fractional Factorial Designs

LEARNING OBJECTIVES

After completing this chapter, you will be able to:

1. Plan, conduct, analyze, and interpret experiments involving a three-level factorial design.

2. Understand the principle of blocking and confounding in three-level factorial designs.

3. Plan, conduct, analyze, and interpret experiments involving a 3^{k-p} fractional factorial design.

4. Understand how the alias structure of a 3^{k-p} fractional factorial design is determined and the concept of partial aliasing in these designs.

5. Understand the basics of non-regular fractional factorial designs.

6. Plan, conduct, analyze, and interpret experiments involving factors with mixed numbers of levels.

KEY CONCEPTS AND IDEAS

1. Components of interaction

2. Aliasing and partial aliasing

3. Quantitative and qualitative factors

4. Blocking and confounding

5. Defining relation

6. Design generator

Exercises

9.1. The effects of developer strength (A) and developer time (B) on the density of photographic plate film are being studied. Three strengths and three times are used, and four replicates of a 3^2 factorial experiment are run. The data from this experiment follow. Analyze the data using the standard methods for factorial experiments.

Developer Strength	Development Time (minutes)					
	10		14		18	
1	0	2	1	3	2	5
	5	4	4	2	4	6
2	4	6	6	8	9	10
	7	5	7	7	8	5
3	7	10	10	10	12	10
	8	7	8	7	9	8

Design-Expert Output

Response: Data
    ANOVA for Selected Factorial Model
Analysis of variance table [Partial sum of squares]

Source	Sum of Squares	DF	Mean Square	F Value	Prob > F	
Model	224.22	8	28.03	10.66	< 0.0001	significant
A	198.22	2	99.11	37.69	< 0.0001	
B	22.72	2	11.36	4.32	0.0236	
AB	3.28	4	0.82	0.31	0.8677	
Pure Error	71.00	27	2.63			
Cor Total	295.22	35				

The Model F-value of 10.66 implies the model is significant. There is only a 0.01% chance that a "Model F-Value" this large could occur due to noise.

Values of "Prob > F" less than 0.0500 indicate model terms are significant. In this case A, B are significant model terms.

Strength and time are significant. The quadratic and interaction effects are not significant. By treating both A and B as numerical factors, the analysis can be performed as follows:

Design-Expert Output

Response: Data
    ANOVA for Selected Factorial Model
Analysis of variance table [Partial sum of squares]

Source	Sum of Squares	DF	Mean Square	F Value	Prob > F	
Model	214.71	2	107.35	44.00	< 0.0001	significant
A	192.67	1	192.67	78.97	< 0.0001	
B	22.04	1	22.04	9.03	0.0050	
Residual	80.51	33	2.44			
Lack of Fit	9.51	6	1.59	0.60	0.7255	not significant
Pure Error	71.00	27	2.63			
Cor Total	295.22	35				

The Model F-value of 44.00 implies the model is significant. There is only a 0.01% chance that a "Model F-Value" this large could occur due to noise.

9.3. An experiment was performed to study the effect of three different types of 32-ounce bottles (A) and three different shelf types (B) – smooth permanent shelves, end-aisle displays with grilled shelves, and beverage coolers – on the time it takes to stock ten 12-bottle cases on the shelves. Three workers (factor C) were employed in this experiment, and two replicates of a 3^3 factorial design were run. The observed time data are shown in the following table. Analyze the data and draw conclusions.

Worker	Bottle Type	Replicate I Permanent	EndAisle	Cooler	Replicate II Permanent	EndAisle	Cooler
1	Plastic	3.45	4.14	5.80	3.36	4.19	5.23
	28-mm glass	4.07	4.38	5.48	3.52	4.26	4.85
	38-mm glass	4.20	4.26	5.67	3.68	4.37	5.58
2	Plastic	4.80	5.22	6.21	4.40	4.70	5.88
	28-mm glass	4.52	5.15	6.25	4.44	4.65	6.20
	38-mm glass	4.96	5.17	6.03	4.39	4.75	6.38
3	Plastic	4.08	3.94	5.14	3.65	4.08	4.49
	28-mm glass	4.30	4.53	4.99	4.04	4.08	4.59
	38-mm glass	4.17	4.86	4.85	3.88	4.48	4.90

Design-Expert Output

Response: Time
     ANOVA for Selected Factorial Model
    Analysis of variance table [Partial sum of squares]

Source	Sum of Squares	DF	Mean Square	F Value	Prob > F	
Model	28.28	26	1.09	14.50	< 0.0001	significant
A	*0.41*	*2*	*0.21*	*2.74*	*0.0828*	
B	*17.75*	*2*	*8.88*	*118.34*	*< 0.0001*	
C	*7.66*	*2*	*3.83*	*51.09*	*< 0.0001*	
AB	*0.12*	*4*	*0.029*	*0.39*	*0.8163*	
AC	*0.11*	*4*	*0.027*	*0.36*	*0.8319*	
BC	*1.68*	*4*	*0.42*	*5.60*	*0.0021*	
ABC	*0.55*	*8*	*0.069*	*0.92*	*0.5145*	
Pure Error	2.03	27	0.075			
Cor Total	30.31	53				

The Model F-value of 14.50 implies the model is significant. There is only a 0.01% chance that a "Model F-Value" this large could occur due to noise.

Values of "Prob > F" less than 0.0500 indicate model terms are significant.
In this case B, C, BC are significant model terms.

Factors B and C, shelf type and worker, and the BC interaction are significant. For the shortest stocking time, independent of the worker who is stocking the shelves, choose the permanent shelves. This can easily be seen in the interaction plot below.

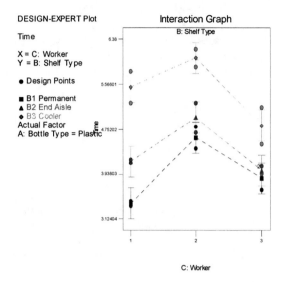

9.9. Consider the data from the first replicate of Problem 9.3. Assuming that not all 27 observations could be run on the same day, set up a design for conducting the experiment over three days with AB^2C confounded with blocks. Analyze the data.

	Block 1			Block 2			Block 3	
000	=	3.45	100	=	4.07	200	=	4.20
110	=	4.38	210	=	4.26	010	=	4.14
011	=	5.22	111	=	5.15	211	=	5.17
102	=	4.30	202	=	4.17	002	=	4.08
201	=	4.96	001	=	4.80	101	=	4.52
212	=	4.86	012	=	3.94	112	=	4.53
121	=	6.25	221	=	6.03	021	=	6.21
022	=	5.14	122	=	4.99	222	=	4.85
220	=	5.67	020	=	5.80	120	=	5.48
Totals	=	44.23			43.21			43.18

The analysis of variance below identifies factors B and C as significant.

Design-Expert Output

Response: Time						
ANOVA for Selected Factorial Model						
Analysis of variance table [Partial sum of squares]						
	Sum of		**Mean**	**F**		
Source	**Squares**	**DF**	**Square**	**Value**	**Prob > F**	
Block	0.079	2	0.040			
Model	13.57	18	0.75	10.49	0.0040	significant
A	*0.11*	*2*	*0.055*	*0.77*	*0.5052*	
B	*8.42*	*2*	*4.21*	*58.63*	*0.0001*	
C	*3.81*	*2*	*1.91*	*26.53*	*0.0010*	
AB	*0.30*	*4*	*0.075*	*1.04*	*0.4573*	
AC	*0.11*	*4*	*0.028*	*0.39*	*0.8105*	
BC	*0.81*	*4*	*0.20*	*2.83*	*0.1234*	
Residual	0.43	6	0.072			
Cor Total	14.08	26				

The Model F-value of 10.49 implies the model is significant. There is only
a 0.40% chance that a "Model F-Value" this large could occur due to noise.

Values of "Prob > F" less than 0.0500 indicate model terms are significant.
In this case B, C are significant model terms.

9.11. Consider the data in Problem 9-3. If ABC is confounded in replicate I and ABC^2 is confounded in replicate II, perform the analysis of variance.

$$L_1 = X_1 + X_2 + X_3 \qquad\qquad L_2 = X_1 + X_2 + 2X_2$$

Block 1		Block 2		Block 3		Block 1		Block 2		Block 3	
000 =	3.45	001 =	4.80	002 =	4.08	000 =	3.36	100 =	3.52	200 =	3.68
111 =	5.15	112 =	4.53	110 =	4.38	101 =	4.44	201 =	4.39	001 =	4.40
222 =	4.85	220 =	5.67	221 =	6.03	011 =	4.70	111 =	4.65	211 =	4.75
120 =	5.48	121 =	6.25	122 =	4.99	221 =	6.38	021 =	5.88	121 =	6.20
102 =	4.30	100 =	4.07	101 =	4.52	202 =	3.88	002 =	3.65	102 =	4.04
210 =	4.26	211 =	5.17	212 =	4.86	022 =	4.49	122 =	4.59	222 =	4.90
201 =	4.96	202 =	4.17	200 =	4.20	120 =	4.85	220 =	5.58	020 =	5.23
012 =	3.94	010 =	4.14	011 =	5.22	210 =	4.37	010 =	4.19	110 =	4.26
021 =	6.21	022 =	5.14	020 =	5.80	112 =	4.08	212 =	4.48	012 =	4.08

The sums of squares for A, B, C, AB, AC, and BC are calculated as usual. The only sums of squares presenting difficulties with calculations are the four components of the ABC interaction (ABC, ABC^2, AB^2C, and AB^2C^2). ABC is computed using replicate I and ABC^2 is computed using replicate II. AB^2C and AB^2C^2 are computed using data from both replicates.

We will show how to calculate AB^2C and AB^2C^2 from both replicates. Form a two-way table of $A \times B$ at each level of C. Find the I(AB) and J(AB) totals for each third of the $A \times B$ table.

C	B	A 0	A 1	A 2	I	J
	0	6.81	7.59	7.88	26.70	27.55
0	1	8.33	8.64	8.63	27.25	27.17
	2	11.03	10.33	11.25	26.54	25.77
	0	9.20	8.96	9.35	31.41	31.24
1	1	9.92	9.80	9.92	30.97	31.29
	2	12.09	12.45	12.41	31.72	31.57
	0	7.73	8.34	8.05	26.09	26.29
2	1	8.02	8.61	9.34	27.31	26.11
	2	9.63	9.58	9.75	25.65	26.65

The I and J components for each third of the above table are used to form a new table of diagonal totals.

C	I(AB)			J(AB)		
0	26.70	27.25	26.54	27.55	27.17	25.77
1	31.41	30.97	31.72	31.24	31.29	31.57
2	26.09	27.31	25.65	26.29	26.11	26.65

	I Totals:		*I* Totals:	
	85.06, 85.26, 83.32		85.49, 85.03, 83.12	
	J Totals:		*J* Totals:	
	85.73, 83.60, 84.31		83.35, 8 5.06, 85.23	

Now, $AB^2C^2 = I[C \times I(AB)] = \dfrac{(85.06)^2 + (85.26)^2 + (83.32)^2}{18} - \dfrac{(253.64)^2}{54} = 0.1265$

and, $AB^2C = J[C \times I(AB)] = \dfrac{(85.73)^2 + (83.60)^2 + (84.31)^2}{18} - \dfrac{(253.64)^2}{54} = 0.1307$

If it were necessary, we could find ABC^2 as $ABC^2 = I[C \times J(AB)]$ and ABC as $J[C \times J(AB)]$. However, these components must be computed using the data from the appropriate replicate.

The analysis of variance table:

Source	SS	DF	MS	F_0
Replicates	1.06696	1		
Blocks within Replicates	0.2038	4		
A	0.4104	2	0.2052	5.02
B	17.7514	2	8.8757	217.0
C	7.6631	2	3.8316	93.68
AB	0.1161	4	0.0290	<1
AC	0.1093	4	0.0273	<1
BC	1.6790	4	0.4198	10.26
ABC (rep I)	0.0452	2	0.0226	<1
ABC^2 (rep II)	0.1020	2	0.0510	1.25
AB^2C	0.1307	2	0.0754	1.60
AB^2C^2	0.1265	2	0.0633	1.55
Error	0.8998	22	0.0409	
Total	30.3069	53		

CHAPTER **10**

Fitting Regression Models

LEARNING OBJECTIVES

After completing this chapter, you will be able to:

1. Estimate the parameters in a linear regression model using the method of least squares.

2. Understand the relationship between a regression model and the 2k factorial design.

3. Test for significance of regression in a linear regression model.

4. Test hypotheses about the individual coefficients in a linear regression model.

5. Construct confidence intervals on the regression coefficients in a linear regression model.

6. Construct confidence intervals on the mean response and prediction intervals on future response observations.

7. Analyze residuals to investigate model adequacy.

KEY CONCEPTS AND IDEAS

1. Linear regression model
2. Least squares estimation of model parameters
3. Test for significance of regression
4. Extra sum of squares method
5. Confidence interval on the mean response
6. Prediction interval on new response values
7. Standardized residuals
8. Studentized residuals
9. PRESS statistic
10. Leverage and influence
11. Lack of fit

Exercises

10.1. The tensile strength of a paper product is related to the amount of hardwood in the pulp. Ten samples are produced in the pilot plant, and the data obtained are shown in the following table.

Strength	Percent Hardwood	Strength	Percent Hardwood
160	10	181	20
171	15	188	25
175	15	193	25
182	20	195	28
184	20	200	30

(a) Fit a linear regression model relating strength to percent hardwood.

MINITAB Output

```
Regression Analysis: Strength versus Hardwood

The regression equation is
Strength = 144 + 1.88 Hardwood

Predictor        Coef     SE Coef          T         P
Constant      143.824       2.522      57.04     0.000
Hardwood       1.8786      0.1165      16.12     0.000

S = 2.203              R-Sq = 97.0%           R-Sq(adj) = 96.6%
PRESS = 66.2665        R-Sq(pred) = 94.91%
```

Regression Plot

Strength = 143.824 + 1.87864 Hardwood

S = 2.20320   R-Sq = 97.0 %   R-Sq(adj) = 96.6 %

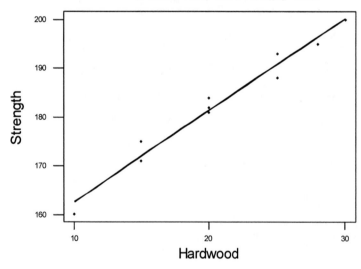

(b) Test the model in part (a) for significance of regression.

MINITAB Output

```
Analysis of Variance

Source            DF         SS         MS         F          P
Regression         1     1262.1     1262.1     260.00     0.000
Residual Error     8       38.8        4.9
  Lack of Fit      4       13.7        3.4       0.54     0.716
  Pure Error       4       25.2        6.3
Total              9     1300.9

3 rows with no replicates

No evidence of lack of fit (P > 0.1)
```

(c) Find a 95 percent confidence interval on the parameter β_1.

The 95 percent confidence interval is:

$$\hat{\beta}_1 - t_{\alpha/2,n-p}se\left(\hat{\beta}_1\right) \leq \beta_1 \leq \hat{\beta}_1 + t_{\alpha/2,n-p}se\left(\hat{\beta}_1\right)$$
$$1.8786 - 2.3060(0.1165) \leq \beta_1 \leq 1.8786 + 2.3060(0.1165)$$
$$1.6900 \leq \beta_1 \leq 2.1473$$

10.7. The brake horsepower developed by an automobile engine on a dynomometer is thought to be a function of the engine speed in revolutions per minute (rpm), the road octane number of the fuel, and the engine compression. An experiment is run in the laboratory and the data that follow are collected.

Brake Horsepower	rpm	Road Octane Number	Compression
225	2000	90	100
212	1800	94	95
229	2400	88	110
222	1900	91	96
219	1600	86	100
278	2500	96	110
246	3000	94	98
237	3200	90	100
233	2800	88	105
224	3400	86	97
223	1800	90	100
230	2500	89	104

(a) Fit a multiple regression model to these data.

MINITAB Output

```
Regression Analysis: Horsepower versus rpm, Octane, Compression

The regression equation is
Horsepower = - 266 + 0.0107 rpm + 3.13 Octane + 1.87 Compression

Predictor        Coef     SE Coef         T       P       VIF
Constant      -266.03       92.67     -2.87   0.021
rpm          0.010713     0.004483      2.39   0.044       1.0
Octane         3.1348       0.8444      3.71   0.006       1.0
Compress       1.8674       0.5345      3.49   0.008       1.0

S = 8.812                R-Sq = 80.7%            R-Sq(adj) = 73.4%
PRESS = 2494.05          R-Sq(pred) = 22.33%
```

(b) Test for significance of regression. What conclusions can you draw?

MINITAB Output

```
Analysis of Variance

Source             DF          SS         MS        F       P
Regression          3     2589.73     863.24    11.12   0.003
Residual Error      8      621.27      77.66
Total              11     3211.00
r No replicates. Cannot do pure error test.

Source       DF      Seq SS
rpm           1      509.35
Octane        1     1132.56
Compress      1      947.83

Lack of fit test
Possible interactions with variable Octane (P-Value = 0.028)
Possible lack of fit at outer X-values      (P-Value = 0.000)
Overall lack of fit test is significant at P = 0.000
```

(c) Based on t tests, do you need all three regressor variables in the model?

Yes, all of the regressor variables are important.

10.9. The yield of a chemical process is related to the concentration of the reactant and the operating temperature. An experiment has been conducted with the following results.

Yield	Concentration	Temperature
81	1.00	150
89	1.00	180
83	2.00	150
91	2.00	180
79	1.00	150
87	1.00	180
84	2.00	150
90	2.00	180

(a) Suppose we wish to fit a main effects model to this data. Set up the **X'X** matrix using the data exactly as it appears in the table.

$$
\begin{bmatrix} 1 & 1 & 1 & 1 & 1 & 1 & 1 & 1 \\ 1.00 & 1.00 & 2.00 & 2.00 & 1.00 & 1.00 & 2.00 & 2.00 \\ 150 & 180 & 150 & 180 & 150 & 180 & 150 & 180 \end{bmatrix}
\begin{bmatrix} 1 & 1.00 & 150 \\ 1 & 1.00 & 180 \\ 1 & 2.00 & 150 \\ 1 & 2.00 & 180 \\ 1 & 1.00 & 150 \\ 1 & 1.00 & 180 \\ 1 & 2.00 & 150 \\ 1 & 2.00 & 180 \end{bmatrix}
=
\begin{bmatrix} 8 & 12 & 1320 \\ 12 & 20 & 1980 \\ 1320 & 1980 & 219600 \end{bmatrix}
$$

(b) Is the matrix you obtained in part (a) diagonal? Discuss your response.

The **X'X** is not diagonal, even though an orthogonal design has been used. The reason is that we have worked with the natural factor levels, not the orthogonally coded variables.

(c) Suppose we write our model in terms of the "usual" coded variables $x_1 = \dfrac{Conc - 1.5}{0.5}$, $x_2 = \dfrac{Temp - 165}{15}$

Set up the **X'X** matrix for the model in terms of these coded variables. Is this matrix diagonal? Discuss your response.

$$
\begin{bmatrix} 1 & 1 & 1 & 1 & 1 & 1 & 1 & 1 \\ -1 & -1 & 1 & 1 & -1 & -1 & 1 & 1 \\ -1 & 1 & -1 & 1 & -1 & 1 & -1 & 1 \end{bmatrix}
\begin{bmatrix} 1 & -1 & -1 \\ 1 & -1 & 1 \\ 1 & 1 & -1 \\ 1 & 1 & 1 \\ 1 & -1 & -1 \\ 1 & -1 & 1 \\ 1 & 1 & -1 \\ 1 & 1 & 1 \end{bmatrix}
=
\begin{bmatrix} 8 & 0 & 0 \\ 0 & 8 & 0 \\ 0 & 0 & 8 \end{bmatrix}
$$

The **X'X** matrix is diagonal because we have used the orthogonally coded variables.

(d) Define a new set of coded variables $x_1 = \dfrac{Conc - 1.0}{1.0}$, $x_2 = \dfrac{Temp - 150}{30}$

Set up the **X'X** matrix for the model in terms of this set of coded variables. Is this matrix diagonal? Discuss your response.

$$
\begin{bmatrix} 1 & 1 & 1 & 1 & 1 & 1 & 1 & 1 \\ 0 & 0 & 1 & 1 & 0 & 0 & 1 & 1 \\ 0 & 1 & 0 & 1 & 0 & 1 & 0 & 1 \end{bmatrix}
\begin{bmatrix} 1 & 0 & 0 \\ 1 & 0 & 1 \\ 1 & 1 & 0 \\ 1 & 1 & 1 \\ 1 & 0 & 0 \\ 1 & 0 & 1 \\ 1 & 1 & 0 \\ 1 & 1 & 1 \end{bmatrix}
=
\begin{bmatrix} 8 & 4 & 4 \\ 4 & 4 & 2 \\ 4 & 2 & 4 \end{bmatrix}
$$

The **X'X** is not diagonal, even though an orthogonal design has been used. The reason is that we have not used orthogonally coded variables.

(e) Summarize what you have learned from this problem about coding the variables.

If the design is orthogonal, use the orthogonal coding. This not only makes the analysis somewhat easier, but it also results in model coefficients that are easier to interpret because they are both dimensionless and uncorrelated.

CHAPTER 11

Response Surface Methods and Designs

LEARNING OBJECTIVES

After completing this chapter, you will be able to:

1. Use the method of steepest ascent to find a region likely to contain the optimum operating conditions for a process.

2. Fit and analyze a second-order response surface model.

3. Simultaneously optimize several responses.

4. Select an appropriate response surface design for fitting either a first-order or a second-order model.

5. Conduct experiments involving mixtures.

6. Fit and interpret a mixture model and determine the optimum formulation.

7. Apply evolutionary operation.

KEY CONCEPTS AND IDEAS

1. Response surface models
2. Method of steepest ascent
3. Ridge systems
4. Multiple response optimization
5. Desirability function
6. Orthogonal first-order design
7. Central composite design

8. Box-Behnken design
9. Rotatable design
10. Face centered cube
11. Computer-generated designs
12. D and G optimal designs
13. Mixture experiments and designs
14. EVOP

Exercises

11.1. A chemical plant produces oxygen by liquefying air and separating it into its component gases by fractional distillation. The purity of the oxygen is a function of the main condenser temperature and the pressure ratio between the upper and lower columns. Current operating conditions are temperature $(\xi_1) = -220°C$ and pressure ratio $(\xi_2) = 1.2$. Using the following data, find the path of steepest ascent.

Temperature (x_1)	Pressure Ratio (x_2)	Purity
-225	1.1	82.8
-225	1.3	83.5
-215	1.1	84.7
-215	1.3	85.0
-220	1.2	84.1
-220	1.2	84.5
-220	1.2	83.9
-220	1.2	84.3

Design-Expert Output

Response: Purity
ANOVA for Selected Factorial Model
Analysis of variance table [Partial sum of squares]

Source	Sum of Squares	DF	Mean Square	F Value	Prob > F	
Model	3.14	2	1.57	26.17	0.0050	significant
A	2.89	1	2.89	48.17	0.0023	
B	0.25	1	0.25	4.17	0.1108	
Curvature	0.080	1	0.080	1.33	0.3125	not significant
Residual	0.24	4	0.060			
Lack of Fit	0.040	1	0.040	0.60	0.4950	not significant
Pure Error	0.20	3	0.067			
Cor Total	3.46	7				

The Model F-value of 26.17 implies the model is significant. There is only a 0.50% chance that a "Model F-Value" this large could occur due to noise.

Std. Dev.	0.24	R-Squared	0.9290	
Mean	84.10	Adj R-Squared	0.8935	
C.V.	0.29	Pred R-Squared	0.7123	
PRESS	1.00	Adeq Precision	12.702	

Factor	Coefficient Estimate	DF	Standard Error	95% CI Low	95% CI High	VIF
Intercept	84.00	1	0.12	83.66	84.34	
A-Temperature	0.85	1	0.12	0.51	1.19	1.00
B-Pressure Ratio	0.25	1	0.12	-0.090	0.59	1.00
Center Point	0.20	1	0.17	-0.28	0.68	1.00

Final Equation in Terms of Coded Factors:

        Purity =

        +84.00

        +0.85   * A

        +0.25   * B

Final Equation in Terms of Actual Factors:

        Purity =

   +118.40000

    +0.17000   * Temperature

    +2.50000   * Pressure Ratio

From the computer output, use the model $\hat{y} = 84 + 0.85x_1 + 0.25x_2$ as the equation for steepest ascent. Suppose we use a one degree change in temperature as the basic step size. Thus, the path of steepest ascent passes through the point (x_1=0, x_2=0) and has a slope 0.25/0.85. In the coded variables, one degree of temperature is equivalent to a step of $\Delta x_1 = 1/5 = 0.2$. Thus, $\Delta x_2 = (0.25/0.85)0.2 = 0.059$. The path of steepest ascent is:

	Coded	Variables	Natural	Variables
	x_1	x_2	ξ_1	ξ_2
Origin	0	0	-220	1.2
Δ	0.2	0.059	1	0.0059
Origin + Δ	0.2	0.059	-219	1.2059
Origin +5 Δ	1.0	0.295	-215	1.2295
Origin +7 Δ	1.40	0.413	-213	1.2413

11.8. The data shown in Table P11.2 were collected in an experiment to optimize crystal growth as a function of three variables x_1, x_2, and x_3. Large values of y (yield in grams) are desirable. Fit a second order model and analyze the fitted surface. Under what set of conditions is maximum growth achieved?

Table P11.2

x_1	x_2	x_3	y
-1	-1	-1	66
-1	-1	1	70
-1	1	-1	78
-1	1	1	60
1	-1	-1	80
1	-1	1	70
1	1	-1	100
1	1	1	75
-1.682	0	0	100
1.682	0	0	80
0	-1.682	0	68
0	1.682	0	63
0	0	-1.682	65
0	0	1.682	82
0	0	0	113
0	0	0	100
0	0	0	118
0	0	0	88
0	0	0	100
0	0	0	85

Design-Expert Output

Response: Yield
    ANOVA for Response Surface Quadratic Model
Analysis of variance table [Partial sum of squares]

Source	Sum of Squares	DF	Mean Square	F Value	Prob > F	
Model	3662.00	9	406.89	2.19	0.1194	not significant
A	22.08	1	22.08	0.12	0.7377	
B	25.31	1	25.31	0.14	0.7200	
C	30.50	1	30.50	0.16	0.6941	
A^2	204.55	1	204.55	1.10	0.3191	
B^2	2226.45	1	2226.45	11.96	0.0061	
C^2	1328.46	1	1328.46	7.14	0.0234	
AB	66.12	1	66.12	0.36	0.5644	
AC	55.13	1	55.13	0.30	0.5982	
BC	171.13	1	171.13	0.92	0.3602	
Residual	1860.95	10	186.09			
Lack of Fit	1001.61	5	200.32	1.17	0.4353	not significant
Pure Error	859.33	5	171.87			
Cor Total	5522.95	19				

The "Model F-value" of 2.19 implies the model is not significant relative to the noise. There is a 11.94 % chance that a "Model F-value" this large could occur due to noise.

Std. Dev.	13.64	R-Squared	0.6631	
Mean	83.05	Adj R-Squared	0.3598	
C.V.	16.43	Pred R-Squared	-0.6034	
PRESS	8855.23	Adeq Precision	3.882	

Factor	Coefficient Estimate	DF	Standard Error	95% CI Low	95% CI High	VIF
Intercept	100.67	1	5.56	88.27	113.06	
A-x1	1.27	1	3.69	-6.95	9.50	1.00
B-x2	1.36	1	3.69	-6.86	9.59	1.00
C-x3	-1.49	1	3.69	-9.72	6.73	1.00
A^2	-3.77	1	3.59	-11.77	4.24	1.02
B^2	-12.43	1	3.59	-20.44	-4.42	1.02
C^2	-9.60	1	3.59	-17.61	-1.59	1.02
AB	2.87	1	4.82	-7.87	13.62	1.00
AC	-2.63	1	4.82	-13.37	8.12	1.00
BC	-4.63	1	4.82	-15.37	6.12	1.00

Final Equation in Terms of Coded Factors:

$$
\begin{aligned}
\text{Yield} =\ & +100.67 \\
& +1.27 * A \\
& +1.36 * B \\
& -1.49 * C \\
& -3.77 * A^2 \\
& -12.43 * B^2 \\
& -9.60 * C^2 \\
& +2.87 * A * B \\
& -2.63 * A * C \\
& -4.63 * B * C
\end{aligned}
$$

Final Equation in Terms of Actual Factors:

$$\begin{aligned}
\text{Yield} = \\
+100.66609 \\
+1.27146 \ * x1 \\
+1.36130 \ * x2 \\
-1.49445 \ * x3 \\
-3.76749 \ * x1^2 \\
-12.42955 \ * x2^2 \\
-9.60113 \ * x3^2 \\
+2.87500 \ * x1 * x2 \\
-2.62500 \ * x1 * x3 \\
-4.62500 \ * x2 * x3
\end{aligned}$$

There are so many nonsignificant terms in this model that we should consider eliminating some of them. A reasonable reduced model is shown in the following Design-Expert output table.

Design-Expert Output

Response: Yield
ANOVA for Response Surface Reduced Quadratic Model
Analysis of variance table [Partial sum of squares]

Source	Sum of Squares	DF	Mean Square	F Value	Prob > F	
Model	3143.00	4	785.75	4.95	0.0095	significant
B	25.31	1	25.31	0.16	0.6952	
C	30.50	1	30.50	0.19	0.6673	
B2	2115.31	1	2115.31	13.33	0.0024	
C2	1239.17	1	1239.17	7.81	0.0136	
Residual	2379.95	15	158.66			
Lack of Fit	1520.62	10	152.06	0.88	0.5953	not significant
Pure Error	859.33	5	171.87			
Cor Total	5522.95	19				

The Model F-value of 4.95 implies the model is significant. There is only a 0.95% chance that a "Model F-Value" this large could occur due to noise.

Std. Dev.	12.60	R-Squared	0.5691	
Mean	83.05	Adj R-Squared	0.4542	
C.V.	15.17	Pred R-Squared	0.1426	
PRESS	4735.52	Adeq Precision	5.778	

Factor	Coefficient Estimate	DF	Standard Error	95% CI Low	95% CI High	VIF
Intercept	97.58	1	4.36	88.29	106.88	
B-x2	1.36	1	3.41	-5.90	8.63	1.00
C-x3	-1.49	1	3.41	-8.76	5.77	1.00
B2	-12.06	1	3.30	-19.09	-5.02	1.01
C2	-9.23	1	3.30	-16.26	-2.19	1.01

Final Equation in Terms of Coded Factors:

$$\begin{aligned}
\text{Yield} = \\
+97.58 \\
+1.36 \ * B \\
-1.49 \ * C \\
-12.06 \ * B^2 \\
-9.23 \ * C^2
\end{aligned}$$

Final Equation in Terms of Actual Factors:

$$Yield =$$
$$+97.58260$$
$$+1.36130 \quad * x2$$
$$-1.49445 \quad * x3$$
$$-12.05546 \quad * x2^2$$
$$-9.22703 \quad * x3^2$$

The contour plot identifies a maximum near the center of the design space.

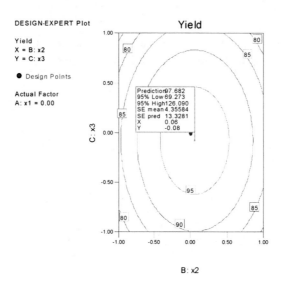

11.12. Consider the three-variable central composite design shown in Table P11.6. Analyze the data and draw conclusions, assuming that we wish to maximize conversion (y_1) with activity (y_2) between 55 and 60.

Table P11.6

Run	Time (min)	Temperature (°C)	Catalyst (%)	Conversion (%) y_1	Activity y_2
1	-1.000	-1.000	-1.000	74.00	53.20
2	1.000	-1.000	-1.000	51.00	62.90
3	-1.000	1.000	-1.000	88.00	53.40
4	1.000	1.000	-1.000	70.00	62.60
5	-1.000	-1.000	1.000	71.00	57.30
6	1.000	-1.000	1.000	90.00	67.90
7	-1.000	1.000	1.000	66.00	59.80
8	1.000	1.000	1.000	97.00	67.80
9	0.000	0.000	0.000	81.00	59.20
10	0.000	0.000	0.000	75.00	60.40
11	0.000	0.000	0.000	76.00	59.10
12	0.000	0.000	0.000	83.00	60.60
13	-1.682	0.000	0.000	76.00	59.10
14	1.682	0.000	0.000	79.00	65.90
15	0.000	-1.682	0.000	85.00	60.00
16	0.000	1.682	0.000	97.00	60.70
17	0.000	0.000	-1.682	55.00	57.40
18	0.000	0.000	1.682	81.00	63.20
19	0.000	0.000	0.000	80.00	60.80
20	0.000	0.000	0.000	91.00	58.90

Quadratic models are developed for the Conversion and Activity response variables as shown below.

Design-Expert Output

Response: Conversion
  ANOVA for Response Surface Quadratic Model
Analysis of variance table [Partial sum of squares]

Source	Sum of Squares	DF	Mean Square	F Value	Prob > F	
Model	2555.73	9	283.97	12.76	0.0002	significant
A	14.44	1	14.44	0.65	0.4391	
B	222.96	1	222.96	10.02	0.0101	
C	525.64	1	525.64	23.63	0.0007	
A²	48.47	1	48.47	2.18	0.1707	
B²	124.48	1	124.48	5.60	0.0396	
C²	388.59	1	388.59	17.47	0.0019	
AB	36.13	1	36.13	1.62	0.2314	
AC	1035.13	1	1035.13	46.53	< 0.0001	
BC	120.12	1	120.12	5.40	0.0425	
Residual	222.47	10	22.25			
Lack of Fit	56.47	5	11.29	0.34	0.8692	not significant
Pure Error	166.00	5	33.20			
Cor Total	287.28	19				

The Model F-value of 12.76 implies the model is significant. There is only
a 0.02% chance that a "Model F-Value" this large could occur due to noise.

Std. Dev.	4.72		R-Squared	0.9199			
Mean	78.30		Adj R-Squared	0.8479			
C.V.	6.02		Pred R-Squared	0.7566			
PRESS	676.22		Adeq Precision	14.239			

Factor	Coefficient Estimate	DF	Standard Error	95% CI Low	95% CI High	VIF
Intercept	81.09	1	1.92	76.81	85.38	
A-Time	1.03	1	1.28	-1.82	3.87	1.00
B-Temperature	4.04	1	1.28	1.20	6.88	1.00
C-Catalyst	6.20	1	1.28	3.36	9.05	1.00
A2	-1.83	1	1.24	-4.60	0.93	1.02
B2	2.94	1	1.24	0.17	5.71	1.02
C2	-5.19	1	1.24	-7.96	-2.42	1.02
AB	2.13	1	1.67	-1.59	5.84	1.00
AC	11.38	1	1.67	7.66	15.09	1.00
BC	-3.87	1	1.67	-7.59	-0.16	1.00

Final Equation in Terms of Coded Factors:

$$
\begin{aligned}
\text{Conversion} = \quad & +81.09 \\
& +1.03 * A \\
& +4.04 * B \\
& +6.20 * C \\
& -1.83 * A^2 \\
& +2.94 * B^2 \\
& -5.19 * C^2 \\
& +2.13 * A * B \\
& +11.38 * A * C \\
& -3.87 * B * C
\end{aligned}
$$

Final Equation in Terms of Actual Factors:

$$
\begin{aligned}
\text{Conversion} = \quad & +81.09128 \\
& +1.02845 * \text{Time} \\
& +4.04057 * \text{Temperature} \\
& +6.20396 * \text{Catalyst} \\
& -1.83398 * \text{Time}^2 \\
& +2.93899 * \text{Temperature}^2 \\
& -5.19274 * \text{Catalyst}^2 \\
& +2.12500 * \text{Time} * \text{Temperature} \\
& +11.37500 * \text{Time} * \text{Catalyst} \\
& -3.87500 * \text{Temperature} * \text{Catalyst}
\end{aligned}
$$

Design-Expert Output

Response: Activity
ANOVA for Response Surface Quadratic Model
Analysis of variance table [Partial sum of squares]

Source	Sum of Squares	DF	Mean Square	F Value	Prob > F	
Model	256.20	9	28.47	9.16	0.0009	significant
A	175.35	1	175.35	56.42	< 0.0001	
B	0.89	1	0.89	0.28	0.6052	
C	67.91	1	67.91	21.85	0.0009	
A2	10.05	1	10.05	3.23	0.1024	
B2	0.081	1	0.081	0.026	0.8753	
C2	0.047	1	0.047	0.015	0.9046	
AB	1.20	1	1.20	0.39	0.5480	
AC	0.011	1	0.011	3.620E-003	0.9532	

BC	*0.78*	*1*	*0.78*	*0.25*	*0.6270*	
Residual	31.08	10	3.11			
Lack of Fit	*27.43*	*5*	*5.49*	*7.51*	*0.0226*	*significant*
Pure Error	*3.65*	*5*	*0.73*			
Cor Total	287.28	19				

The Model F-value of 9.16 implies the model is significant. There is only a 0.09% chance that a "Model F-Value" this large could occur due to noise.

Std. Dev.	1.76	R-Squared	0.8918	
Mean	60.51	Adj R-Squared	0.7945	
C.V.	2.91	Pred R-Squared	0.2536	
PRESS	214.43	Adeq Precision	10.911	

Factor	Coefficient Estimate	DF	Standard Error	95% CI Low	95% CI High	VIF
Intercept	59.85	1	0.72	58.25	61.45	
A-Time	3.58	1	0.48	2.52	4.65	1.00
B-Temperature	0.25	1	0.48	-0.81	1.32	1.00
C-Catalyst	2.23	1	0.48	1.17	3.29	1.00
A^2	0.83	1	0.46	-0.20	1.87	1.02
B^2	0.075	1	0.46	-0.96	1.11	1.02
C^2	0.057	1	0.46	-0.98	1.09	1.02
AB	-0.39	1	0.62	-1.78	1.00	1.00
AC	-0.038	1	0.62	-1.43	1.35	1.00
BC	0.31	1	0.62	-1.08	1.70	1.00

Final Equation in Terms of Coded Factors:

$$
\begin{aligned}
\text{Conversion} =\ & +59.85 \\
& +3.58 * A \\
& +0.25 * B \\
& +2.23 * C \\
& +0.83 * A^2 \\
& +0.075 * B^2 \\
& +0.057 * C^2 \\
& -0.39 * A * B \\
& -0.038 * A * C \\
& +0.31 * B * C
\end{aligned}
$$

Final Equation in Terms of Actual Factors:

$$
\begin{aligned}
\text{Conversion} =\ & +59.84984 \\
& +3.58327 * \text{Time} \\
& +0.25462 * \text{Temperature} \\
& +2.22997 * \text{Catalyst} \\
& +0.83491 * \text{Time}^2 \\
& +0.074772 * \text{Temperature}^2 \\
& +0.057094 * \text{Catalyst}^2 \\
& -0.38750 * \text{Time} * \text{Temperature} \\
& -0.037500 * \text{Time} * \text{Catalyst} \\
& +0.31250 * \text{Temperature} * \text{Catalyst}
\end{aligned}
$$

Because many of the terms are insignificant, a reduced quadratic model was fit and is shown below.

Design-Expert Output

Response: Activity
    ANOVA for Response Surface Quadratic Model
Analysis of variance table [Partial sum of squares]

Source	Sum of Squares	DF	Mean Square	F Value	Prob > F	
Model	253.20	3	84.40	39.63	< 0.0001	significant
A	175.35	1	175.35	82.34	< 0.0001	
C	67.91	1	67.91	31.89	< 0.0001	
A^2	9.94	1	9.94	4.67	0.0463	
Residual	34.07	16	2.13			
Lack of Fit	30.42	11	2.77	3.78	0.0766	not significant
Pure Error	3.65	5	0.73			
Cor Total	287.28	19				

The Model F-value of 39.63 implies the model is significant. There is only
a 0.01% chance that a "Model F-Value" this large could occur due to noise.

Std. Dev.	1.46	R-Squared	0.8814	
Mean	60.51	Adj R-Squared	0.8591	
C.V.	2.41	Pred R-Squared	0.6302	
PRESS	106.24	Adeq Precision	20.447	

Factor	Coefficient Estimate	DF	Standard Error	95% CI Low	95% CI High	VIF
Intercept	59.95	1	0.42	59.06	60.83	
A-Time	3.58	1	0.39	2.75	4.42	1.00
C-Catalyst	2.23	1	0.39	1.39	3.07	1.00
A^2	0.82	1	0.38	0.015	1.63	1.00

Final Equation in Terms of Coded Factors:

    Activity =
    +59.95
    +3.58 * A
    +2.23 * C
    +0.82 * A^2

Final Equation in Terms of Actual Factors:

    Activity =
  +59.94802
    +3.58327 * Time
    +2.22997 * Catalyst
    +0.82300 * $Time^2$

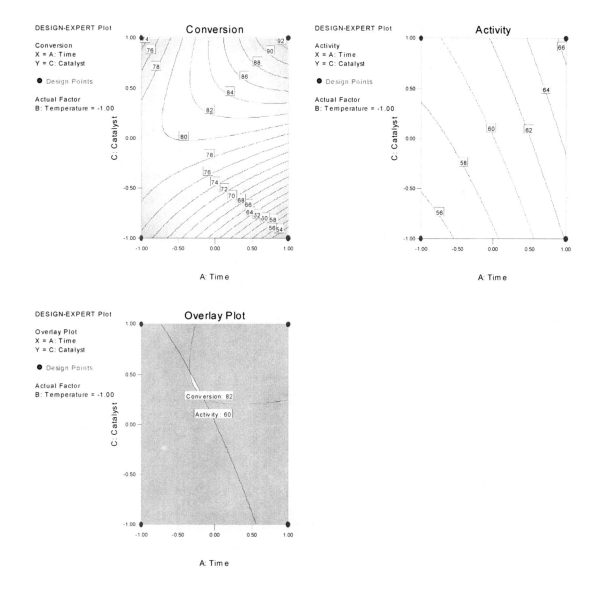

The contour plots visually describe the models while the overlay plot identifies the acceptable region for the process.

11.15. Verify that an orthogonal first-order design is also first-order rotatable.

To show that a first order orthogonal design is also first order rotatable, consider

$$V(\hat{y}) = V(\hat{\beta}_0 + \sum_{i=1}^{k} \hat{\beta}_i x_i) = V(\hat{\beta}_0) + \sum_{i=1}^{k} x_i^2 V(\hat{\beta}_i)$$

since all covariances between $\hat{\beta}_i$ and $\hat{\beta}_j$ are zero, due to design orthogonality. Furthermore, we have:

$$V\left(\hat{\beta}_0\right) = V(\hat{\beta}_1) = V(\hat{\beta}_2) = \ldots = V(\hat{\beta}_k) = \frac{\sigma^2}{n}, \text{ so}$$

$$V(\hat{y}) = \frac{\sigma^2}{n} + \frac{\sigma^2}{n}\sum_{i=1}^{k} x_i^2$$

$$V(\hat{y}) = \frac{\sigma^2}{n}\left(1 + \sum_{i=1}^{k} x_i^2\right)$$

which is a function of distance from the design center (i.e. **x=0**), and not direction. Thus the design must be rotatable. Note that n is, in general, the number of points in the exterior portion of the design. If there are n_c centerpoints, then $V(\hat{\beta}_0) = \dfrac{\sigma^2}{(n + n_c)}$.

11.23. Suppose that you need to design an experiment to fit a quadratic model over the region $-1 \le x_i \le +1$, $i=1, 2$ subject to the constraint $x_1 + x_2 \le 1$. If the constraint is violated, the process will not work properly. You can afford to make no more than n=12 runs. Set up the following designs:

(a) An "inscribed" CCD with center points at $x_1 = x_2 = 0$.

x_1	x_2
-0.5	-0.5
0.5	-0.5
-0.5	0.5
0.5	0.5
-0.707	0
0.707	0
0	-0.707
0	0.707
0	0
0	0
0	0
0	0

(b) An "inscribed" 3^2 factorial with center points at $x_1 = x_2 - 0.25$.

x_1	x_2
-1	-1
-0.25	-1
0.5	-1
-1	-0.25
-0.25	-0.25
0.5	-0.25
-1	0.5
-0.25	0.5
0.5	0.5
-0.25	-0.25
-0.25	-0.25
-0.25	-0.25

(c) A D-optimal design.

x_1	x_2
-1	-1
1	-1
-1	1
1	0
0	1
0	0
-1	0
0	-1
0.5	0.5
-1	-1
1	-1
-1	1

(d) A modified D-optimal design that is identical to the one in part (c), but with all replicate runs at the design center.

x_1	x_2
1	0
0	0
0	1
-1	-1
1	-1
-1	1
-1	0
0	-1
0.5	0.5
0	0
0	0
0	0

(e) Evaluate the $\left|(\mathbf{X'X})^{-1}\right|$ criteria for each design.

	(a)	(b)	(c)	(d)
$\left|(\mathbf{X'X})^{-1}\right|$	0.5	0.007217	0.0001016	0.0002294

(f) Evaluate the D-efficiency for each design relative to the D-optimal design in part (c).

	(a)	(b)	(c)	(d)
D-efficiency	24.25%	49.14%	100.00%	87.31%

(g) Which design would you prefer? Why?

The D-optimal is the preferred design based on the D-efficiency. Note: If the CCD was constructed with the center point at $x_1 = x_2 = -0.25$, then a larger design can be fit within the region. This design is:

x_1	x_2
-1	-1
0.5	-1
-1	0.5
0.5	0.5
-1.664	-0.25
1.164	-0.25
-0.25	-1.664
-0.25	1.164
-0.25	-0.25
-0.25	-0.25
-0.25	-0.25
-0.25	-0.25

For this design, $\left|(\mathbf{X'X})^{-1}\right| = 0.00005248$. Its efficiency relative to the D-optimal design is 111.64%. In other words, it is actually a better design.

11.24. Consider a 2^3 design for fitting a first-order model.

(a) Evaluate the D-criterion $\left|(\mathbf{X'X})^{-1}\right|$ for this design.

$$\left|(\mathbf{X'X})^{-1}\right| = 2.441\text{E-}4$$

(b) Evaluate the A-criterion $tr(\mathbf{X'X})^{-1}$ for this design.

$$tr(\mathbf{X'X})^{-1} = 0.5$$

(c) Find the maximum scaled prediction variance for this design. Is this design G-optimal?

$$v(\mathbf{x}) = \frac{NVar(\hat{y}(\mathbf{x}))}{\sigma^2} = N\mathbf{x}'^{(1)}(\mathbf{X}'\mathbf{X})^{-1}\mathbf{x}^{(1)} = 4$$

Yes, this is a G-optimal design.

11.30. An experimenter wishes to run a three-component mixture experiment. The constraints in the components proportions are as follows:

$$0.2 \le x_1 \le 0.4$$
$$0.1 \le x_2 \le 0.3$$
$$0.4 \le x_3 \le 0.7$$

(a) Set up an experiment to fit a quadratic mixture model. Use $n=14$ runs, with 4 replicates. Use the D-criterion.

Std	x_1	x_2	x_3
1	0.2	0.3	0.5
2	0.3	0.3	0.4
3	0.3	0.15	0.55
4	0.2	0.1	0.7
5	0.4	0.2	0.4
6	0.4	0.1	0.5
7	0.2	0.2	0.6
8	0.275	0.25	0.475
9	0.35	0.175	0.475
10	0.3	0.1	0.6
11	0.2	0.3	0.5
12	0.3	0.3	0.4
13	0.2	0.1	0.7
14	0.4	0.1	0.5

(b) Draw the experimental design region.

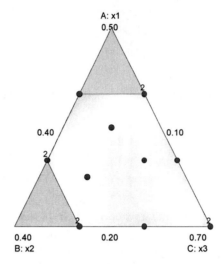

(c) Set up an experiment to fit a quadratic mixture model with $n=12$ runs, assuming that three of these runs are replicated. Use the D-criterion.

Std	x_1	x_2	x_3
1	0.3	0.15	0.55
2	0.2	0.3	0.5
3	0.3	0.3	0.4
4	0.2	0.1	0.7
5	0.4	0.2	0.4
6	0.4	0.1	0.5
7	0.2	0.2	0.6
8	0.275	0.25	0.475
9	0.35	0.175	0.475
10	0.2	0.1	0.7
11	0.4	0.1	0.5
12	0.4	0.2	0.4

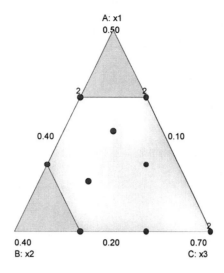

(d) Comment on the two designs you have found.

The design points are the same for both designs except that the edge center on the x_1-x_3 edge is not included in the second design. None of the replicates for either design are in the center of the experimental region. The experimental runs are fairly uniformly spaced in the design region.

11.31. Myers, Montgomery and Anderson-Cook (2009) describe a gasoline blending experiment involving three mixture components. There are no constraints on the mixture proportions, and the following 10-run design is used.

Design Point	x_1	x_2	x_3	y(mpg)
1	1	0	0	24.5, 25.1
2	0	1	0	24.8, 23.9
3	0	0	1	22.7, 23.6
4	½	½	0	25.1
5	½	0	½	24.3
6	0	½	½	23.5
7	1/3	1/3	1/3	24.8, 24.1
8	2/3	1/6	1/6	24.2
9	1/6	2/3	1/6	23.9
10	1/6	1/6	2/3	23.7

(a) What type of design did the experimenters use?

A simplex centroid design was used.

(b) Fit a quadratic mixture model to the data. Is this model adequate?

Design-Expert Output

Response: y
ANOVA for Mixture Quadratic Model
Analysis of variance table [Partial sum of squares]

Source	Sum of Squares	DF	Mean Square	F Value	Prob > F	
Model	4.22	5	0.84	3.90	0.0435	significant
Linear Mixture	3.92	2	1.96	9.06	0.0088	
AB	0.15	1	0.15	0.69	0.4289	
AC	0.081	1	0.081	0.38	0.5569	
BC	0.077	1	0.077	0.36	0.5664	
Residual	1.73	8	0.22			
Lack of Fit	0.50	4	0.12	0.40	0.8003	not significant
Pure Error	1.24	4	0.31			
Cor Total	5.95	13				

The Model F-value of 3.90 implies the model is significant. There is only
a 4.35% chance that a "Model F-Value" this large could occur due to noise.

Std. Dev.	0.47	R-Squared	0.7091	
Mean	24.16	Adj R-Squared	0.5274	
C.V.	1.93	Pred R-Squared	0.1144	
PRESS	5.27	Adeq Precision	5.674	

Component	Coefficient Estimate	DF	Standard Error	95% CI Low	95% CI High
A-x1	24.74	1	0.32	24.00	25.49
B-x2	24.31	1	0.32	23.57	25.05
C-x3	23.18	1	0.32	22.43	23.92
AB	1.51	1	1.82	-2.68	5.70
AC	1.11	1	1.82	-3.08	5.30
BC	-1.09	1	1.82	-5.28	3.10

Final Equation in Terms of Pseudo Components:

$$
\begin{aligned}
y =\ & \\
+24.74\ & * A \\
+24.31\ & * B \\
+23.18\ & * C \\
+1.51\ & * A * B \\
+1.11\ & * A * C \\
-1.09\ & * B * C
\end{aligned}
$$

Final Equation in Terms of Real Components:

$$
\begin{aligned}
y =\ & \\
+24.74432\ & * x1 \\
+24.31098\ & * x2 \\
+23.17765\ & * x3 \\
+1.51364\ & * x1 * x2 \\
+1.11364\ & * x1 * x3 \\
-1.08636\ & * x2 * x3
\end{aligned}
$$

The quadratic terms appear to be insignificant. The analysis below is for the linear mixture model.

Design-Expert Output

Response: y
ANOVA for Mixture Quadratic Model
Analysis of variance table [Partial sum of squares]

Source	Sum of Squares	DF	Mean Square	F Value	Prob > F	
Model	3.92	2	1.96	10.64	0.0027	significant
Linear Mixture	3.92	2	1.96	10.64	0.0027	
Residual	2.03	11	0.18			
Lack of Fit	0.79	7	0.11	0.37	0.8825	not significant
Pure Error	1.24	4	0.31			
Cor Total	5.95	13				

The Model F-value of 10.64 implies the model is significant. There is only
a 0.27% chance that a "Model F-Value" this large could occur due to noise.

Std. Dev.	0.43	R-Squared	0.6591	
Mean	24.16	Adj R-Squared	0.5972	
C.V.	1.78	Pred R-Squared	0.3926	
PRESS	3.62	Adeq Precision	8.751	

Component	Coefficient Estimate	DF	Standard Error	95% CI Low	95% CI High
A-x1	24.93	1	0.25	24.38	25.48
B-x2	24.35	1	0.25	23.80	24.90
C-x3	23.19	1	0.25	22.64	23.74

Component	Effect	DF	Adjusted Std Error	Adjusted Effect=0	Approx t for H0 Prob > |t|
A-x1	1.16	1	0.33	3.49	0.0051
B-x2	0.29	1	0.33	0.87	0.4021
C-x3	-1.45	1	0.33	-4.36	0.0011

Final Equation in Terms of Pseudo Components:

$$y =$$
$$+24.93 * A$$
$$+24.35 * B$$
$$+23.19 * C$$

Final Equation in Terms of Real Components:

$$y =$$
$$+24.93048 * x1$$
$$+24.35048 * x2$$
$$+23.19048 * x3$$

(c) Plot the response surface contours. What blend would you recommend to maximize the miles per gallon?

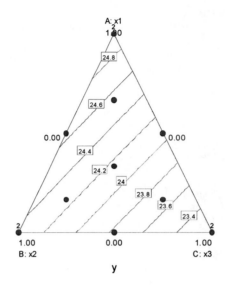

To maximize the miles per gallon, the recommended blend is $x_1 = 1$, $x_2 = 0$, and $x_3 = 0$.

CHAPTER 12

Robust Parameter Design and Process Robustness Studies

LEARNING OBJECTIVES

After completing this chapter, you will be able to:

1. Design and analyze experiments that find the settings of controllable factors that minimize the effects of the uncontrollable or noise factors.

2. Understand the advantages and inefficiencies of Taguchi's crossed array designs.

3. Use response surface methods for process robustness studies or robust product/process design.

KEY CONCEPTS AND IDEAS

1. Robust design

2. Controllable and noise factors

3. Crossed array design

4. Combined array design

5. Propagation of error

Exercises

12.2 Consider the bottle filling experiment in Problem 6.18. Suppose that the percentage of carbonation (A) is a noise variable ($\sigma_z^2 = 1$ in coded units).

(a) Fit the response model to these data. Is there a robust design problem?

The following is the analysis of variance with all terms in the model followed by a reduced model. Because the noise factor A is significant, and the AB interaction is moderately significant, there is a robust design problem.

Design-Expert Output

Response: Fill Deviation						
ANOVA for Response Surface Reduced Cubic Model						
Analysis of variance table [Partial sum of squares]						
	Sum of		Mean	F		
Source	Squares	DF	Square	Value	Prob > F	
Cor Total	300.05	3				
Model	73.00	7	10.43	16.69	0.0003	significant
A	*36.00*	*1*	*36.00*	*57.60*	*< 0.0001*	
B	*20.25*	*1*	*20.25*	*32.40*	*0.0005*	
C	*12.25*	*1*	*12.25*	*19.60*	*0.0022*	
AB	*2.25*	*1*	*2.25*	*3.60*	*0.0943*	
AC	*0.25*	*1*	*0.25*	*0.40*	*0.5447*	
BC	*1.00*	*1*	*1.00*	*1.60*	*0.2415*	
ABC	*1.00*	*1*	*1.00*	*1.60*	*0.2415*	
Pure Error	5.00	8	0.63			
Cor Total	78.00	15				

Based on the above analysis, the AC, BC, and ABC interactions are removed from the model and used as error.

Design-Expert Output

Response: Fill Deviation						
ANOVA for Response Surface Reduced Cubic Model						
Analysis of variance table [Partial sum of squares]						
	Sum of		Mean	F		
Source	Squares	DF	Square	Value	Prob > F	
Model	70.75	4	17.69	26.84	< 0.0001	significant
A	*36.00*	*1*	*36.00*	*54.62*	*< 0.0001*	
B	*20.25*	*1*	*20.25*	*30.72*	*0.0002*	
C	*12.25*	*1*	*12.25*	*18.59*	*0.0012*	
AB	*2.25*	*1*	*2.25*	*3.41*	*0.0917*	
Residual	7.25	11	0.66			
Lack of Fit	2.25	3	0.75	1.20	0.3700	not significant
Pure Error	5.00	8	0.63			
Cor Total	78.00	15				

The Model F-value of 26.84 implies there is a 0.01% chance that a "Model F-Value" this large could occur due to noise.

Std. Dev.	0.81	R-Squared	0.9071
Mean	1.00	Adj R-Squared	0.8733
C.V.	81.18	Pred R-Squared	0.8033
PRESS	15.34	Adeq Precision	15.424

Final Equation in Terms of Coded Factors:

Fill Deviation =
+1.00
+1.50   * A
+1.13   * B
+0.88   * C
+0.38   * A * B

(b) Find the mean model and either the variance model or the POE.

From the final equation shown in the above analysis, the mean model and corresponding contour plot is shown below.

$$E_z\left[y(\mathbf{x},z_1)\right]=1+1.13x_2+0.88x_3$$

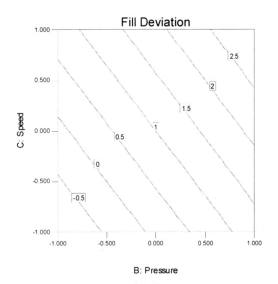

The Contour and 3-D plots of the POE are shown on the next page.

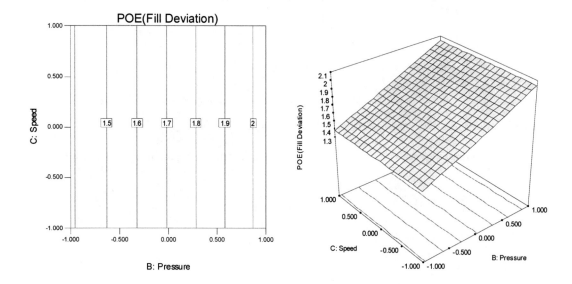

(c) Find a set of conditions that result in mean fill deviation as close to zero as possible with minimum transmitted variance.

The overlay plot below identifies an operating region for pressure and speed that result in a mean fill deviation as close to zero as possible with minimum transmitted variance.

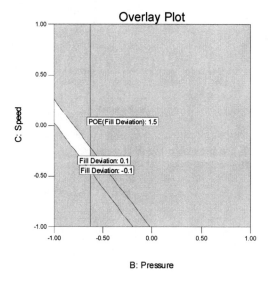

12.8. Consider the experiment in Problem 11.11. Suppose that pressure is a noise variable ($\sigma_z^2 = 1$ in coded units). Fit the response model for the viscosity response. Find a set of conditions that result in viscosity as close as possible to 600 and that minimize the variability transmitted from the noise variable pressure.

Design-Expert Output

Response: Viscosity
     ANOVA for Response Surface Quadratic Model
Analysis of variance table [Partial sum of squares]

Source	Sum of Squares	DF	Mean Square	F Value	Prob > F	
Model	85467.33	6	14244.56	12.12	0.0012	significant
A	703.12	1	703.12	0.60	0.4615	
B	6105.12	1	6105.12	5.19	0.0522	
C	5408.00	1	5408.00	4.60	0.0643	
A2	21736.93	1	21736.93	18.49	0.0026	
C2	5153.80	1	5153.80	4.38	0.0696	
AC	47742.25	1	47742.25	40.61	0.0002	
Residual	9404.00	8	1175.50			
Lack of Fit	7922.00	6	1320.33	1.78	0.4022	not significant
Pure Error	1482.00	2	741.00			
Cor Total	94871.33	14				

The Model F-value of 12.12 implies the model is significant. There is only a 0.12% chance that a "Model F-Value" this large could occur due to noise.

Std. Dev.	34.29	R-Squared	0.9009	
Mean	575.33	Adj R-Squared	0.8265	
C.V.	5.96	Pred R-Squared	0.6279	
PRESS	35301.77	Adeq Precision	11.731	

Final Equation in Terms of Coded Factors:

$$\begin{aligned}
\text{Viscosity} =\ & +636.00 \\
& +9.37 * A \\
& +27.62 * B \\
& -26.00 * C \\
& -76.50 * A^2 \\
& -37.25 * C^2 \\
& +109.25 * A * C
\end{aligned}$$

From the final equation shown in the above analysis, the mean model is shown below.

$$E_z\left[y(\mathbf{x}, z_1)\right] = 636.00 + 9.37x_1 + 27.62x_2 - 26.00x_3 - 76.50x_1^2 - 37.25x_3^2 + 109.25x_1 x_3$$

The corresponding contour and 3-D plots for this model are shown below followed by the POE contour and 3-D plots. Finally, the stacked contour plot is presented, identifying a region with viscosity between 590 and 610 while minimizing the variability transmitted from the noise variable pressure. These conditions are in the region of factor $A = 0.5$ and factor $B = -1$.

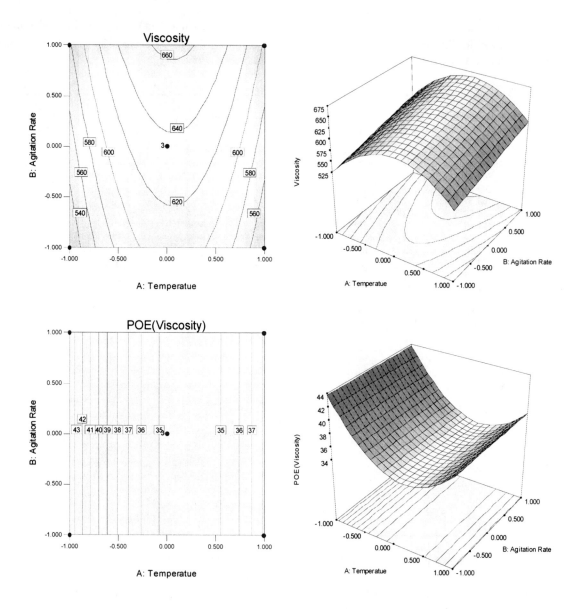

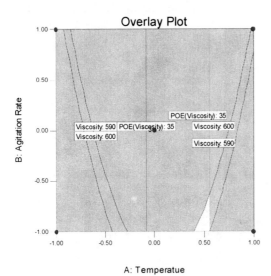

12.10. In an article ("Let's All Beware the Latin Square," *Quality Engineering*, Vol. 1, 1989, pp. 453-465) J.S. Hunter illustrates some of the problems associated with 3^{k-p} fractional factorial designs. Factor A is the amount of ethanol added to a standard fuel and factor B represents the air/fuel ratio. The response variable is carbon monoxide (CO) emission in g/m^2. The design is as follows,

Design				Observations	
A	B	x_1	x_2	y	
0	0	-1	-1	66	62
1	0	0	-1	78	81
2	0	1	-1	90	94
0	1	-1	0	72	67
1	1	0	0	80	81
2	1	1	0	75	78
0	2	-1	1	68	66
1	2	0	1	66	69
2	2	1	1	60	58

Notice that we have used the notation system of 0, 1, and 2 to represent the low, medium, and high levels for the factors. We have also used a "geometric notation" of -1, 0, and 1. Each run in the design is replicated twice.

(a) Verify that the second-order model

$$\hat{y} = 78.5 + 4.5x_1 - 7.0x_2 - 4.5x_1^2 - 4.0x_2^2 - 9.0x_1x_2$$

is a reasonable model for this experiment. Sketch the CO concentration contours in the x_1, x_2 space.

In the computer output that follows, the "coded factors" model is in the -1, 0, +1 scale.

Design-Expert Output

Response: CO Emis
    ANOVA for Response Surface Quadratic Model
Analysis of variance table [Partial sum of squares]

Source	Sum of Squares	DF	Mean Square	F Value	Prob > F	
Model	1624.00	5	324.80	50.95	< 0.0001	significant
A	243.00	1	243.00	38.12	< 0.0001	
B	588.00	1	588.00	92.24	< 0.0001	
A^2	81.00	1	81.00	12.71	0.0039	
B^2	64.00	1	64.00	10.04	0.0081	
AB	648.00	1	648.00	101.65	< 0.0001	
Residual	76.50	12	6.37			
Lack of Fit	30.00	3	10.00	1.94	0.1944	not significant
Pure Error	46.50	9	5.17			
Cor Total	1700.50	17				

The Model F-value of 50.95 implies the model is significant. There is only
a 0.01% chance that a "Model F-Value" this large could occur due to noise.

Std. Dev.	2.52	R-Squared	0.9550	
Mean	72.83	Adj R-Squared	0.9363	
C.V.	3.47	Pred R-Squared	0.9002	
PRESS	169.71	Adeq Precision	21.952	

Factor	Coefficient Estimate	DF	Standard Error	95% CI Low	95% CI High	VIF
Intercept	78.50	1	1.33	75.60	81.40	
A-Ethanol	4.50	1	0.73	2.91	6.09	1.00
B-Air/Fuel Ratio	-7.00	1	0.73	-8.59	-5.41	1.00
A^2	-4.50	1	1.26	-7.25	-1.75	1.00
B^2	-4.00	1	1.26	-6.75	-1.25	1.00
AB	-9.00	1	0.89	-10.94	-7.06	1.00

Final Equation in Terms of Coded Factors:

$$
\begin{aligned}
\text{CO Emis} = \ & +78.50 \\
& +4.50 \ * A \\
& -7.00 \ * B \\
& -4.50 \ * A^2 \\
& -4.00 \ * B^2 \\
& -9.00 \ * A * B
\end{aligned}
$$

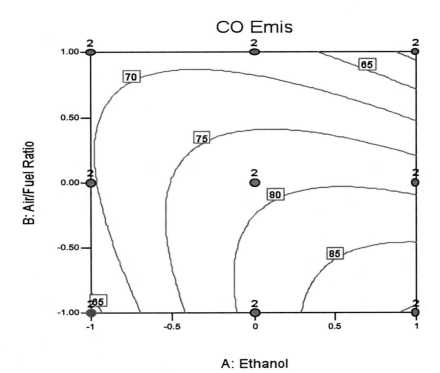

CO Emis

B: Air/Fuel Ratio

A: Ethanol

(b) Now suppose that instead of only two factors, we had used *four* factors in a 3^{4-2} fractional factorial design and obtained *exactly* the same data in part (a). The design would be as follows:

Design								Observations	
A	B	C	D	x_1	x_2	x_3	x_4	y	y
0	0	0	0	-1	-1	-1	-1	66	62
1	0	1	1	0	-1	0	0	78	81
2	0	2	2	+1	-1	+1	+1	90	94
0	1	2	1	-1	0	+1	0	72	67
1	1	0	2	0	0	-1	+1	80	81
2	1	1	0	+1	0	0	-1	75	78
0	2	1	2	-1	+1	0	+1	68	66
1	2	2	0	0	+1	+1	-1	66	69
2	2	0	1	+1	+1	-1	0	60	58

Calculate the marginal averages of the CO response at each level of four factors A, B, C, and D. Construct plots of these marginal averages and interpret the results. Do factors C and D appear to have strong effects? Do these factors *really* have any effect on CO emission? Why is their apparent effect strong?

The marginal averages are shown below. Both factors C and D appear to have an effect on CO emission. This is probably because both C and D are aliased with components of the interaction involving A and B, and there is a strong AB interaction.

Level	A	B	C	D
0	66.83	78.50	67.83	69.33
1	75.83	75.50	74.33	69.33
2	75.83	64.50	76.33	79.83

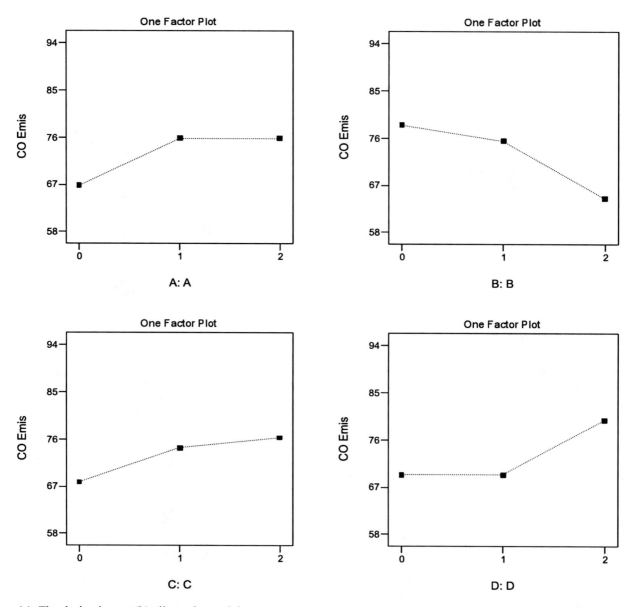

(c) The design in part (b) allows the model

$$y = \beta_0 + \sum_{i=1}^{4} \beta_i x_i + \sum_{i=1}^{4} \beta_{ii} x_i^2 + \varepsilon$$

to be fitted. Suppose that the *true* model is

$$y = \beta_0 + \sum_{i=1}^{4} \beta_i x_i + \sum_{i=1}^{4} \beta_{ii} x_i^2 + \sum \sum_{i<j} \beta_{ij} x_i x_j + \varepsilon$$

Show that if $\hat{\beta}_i$ represents the least squares estimates of the coefficients in the fitted model, then

$$E\left(\hat{\beta}_0\right) = \beta_0 - \beta_{13} - \beta_{14} - \beta_{34}$$
$$E\left(\hat{\beta}_1\right) = \beta_1 - \left(\beta_{23} + \beta_{24}\right)/2$$
$$E\left(\hat{\beta}_2\right) = \beta_2 - \left(\beta_{13} + \beta_{14} + \beta_{34}\right)/2$$
$$E\left(\hat{\beta}_3\right) = \beta_3 - \left(\beta_{12} + \beta_{24}\right)/2$$
$$E\left(\hat{\beta}_4\right) = \beta_4 - \left(\beta_{12} + \beta_{23}\right)/2$$
$$E\left(\hat{\beta}_{11}\right) = \beta_{11} - \left(\beta_{23} - \beta_{24}\right)/2$$
$$E\left(\hat{\beta}_{22}\right) = \beta_{22} + \left(\beta_{13} + \beta_{14} + \beta_{34}\right)/2$$
$$E\left(\hat{\beta}_{33}\right) = \beta_{33} - \left(\beta_{24} - \beta_{12}\right)/2 + \beta_{14}$$
$$E\left(\hat{\beta}_{44}\right) = \beta_{44} - \left(\beta_{12} - \beta_{23}\right)/2 + \beta_{13}$$

Does this help explain the strong effects for factors C and D observed graphically in part (b)?

Let $\mathbf{X}_1 =$

β_0	β_1	β_2	β_3	β_4	β_{11}	β_{22}	β_{33}	β_{44}
1	−1	−1	−1	−1	1	1	1	1
1	0	−1	0	0	0	1	0	0
1	1	−1	1	1	1	1	1	1
1	−1	0	1	0	1	0	1	0
1	0	0	−1	1	0	0	1	1
1	1	0	0	−1	1	0	0	1
1	−1	1	0	1	1	1	0	1
1	0	1	1	−1	0	1	1	1
1	1	1	−1	0	1	1	1	0

and $\mathbf{X}_2 =$

β_{12}	β_{13}	β_{14}	β_{23}	β_{24}	β_{34}
1	1	1	1	1	1
0	0	0	0	0	0
−1	1	1	−1	−1	1
0	−1	0	0	0	0
0	0	0	0	0	−1
0	0	−1	0	0	0
−1	0	−1	0	1	0
0	0	0	1	−1	−1
1	−1	0	−1	0	0

Then, $\mathbf{A} = \left(\mathbf{X}_1'\mathbf{X}_1\right)^{-1}\mathbf{X}_1'\mathbf{X}_2 = \begin{bmatrix} 0 & -1 & -1 & 0 & 0 & -1 \\ 0 & 0 & 0 & -1/2 & -1/2 & 0 \\ 0 & -1/2 & -1/2 & 0 & 0 & -1/2 \\ -1/2 & 0 & 0 & 0 & -1/2 & 0 \\ -1/2 & 0 & 0 & -1/2 & 0 & 0 \\ 0 & 0 & 0 & -1/2 & 1/2 & 1 \\ 0 & 1/2 & 1/2 & 0 & 0 & 1/2 \\ 1/2 & 0 & 1 & 0 & -1/2 & 0 \\ -1/2 & 1 & 0 & 1/2 & 0 & 0 \end{bmatrix}$

$$E\begin{bmatrix} \hat{\beta}_0 \\ \hat{\beta}_1 \\ \hat{\beta}_2 \\ \hat{\beta}_3 \\ \hat{\beta}_4 \\ \hat{\beta}_{11} \\ \hat{\beta}_{22} \\ \hat{\beta}_{33} \\ \hat{\beta}_{44} \end{bmatrix} = \begin{bmatrix} \beta_0 \\ \beta_1 \\ \beta_2 \\ \beta_3 \\ \beta_4 \\ \beta_{11} \\ \beta_{22} \\ \beta_{33} \\ \beta_{44} \end{bmatrix} + \begin{bmatrix} 0 & -1 & -1 & 0 & 0 & -1 \\ 0 & 0 & 0 & -1/2 & -1/2 & 0 \\ 0 & -1/2 & -1/2 & 0 & 0 & -1/2 \\ -1/2 & 0 & 0 & 0 & -1/2 & 0 \\ -1/2 & 0 & 0 & -1/2 & 0 & 0 \\ 0 & 0 & 0 & -1/2 & 1/2 & 1 \\ 0 & 1/2 & 1/2 & 0 & 0 & 1/2 \\ 1/2 & 0 & 1 & 0 & -1/2 & 0 \\ -1/2 & 1 & 0 & 1/2 & 0 & 0 \end{bmatrix} \begin{bmatrix} \beta_{12} \\ \beta_{13} \\ \beta_{14} \\ \beta_{23} \\ \beta_{24} \\ \beta_{34} \end{bmatrix} = \begin{bmatrix} \beta_0 - \beta_{13} - \beta_{14} - \beta_{34} \\ \beta_1 - 1/2\,\beta_{23} - 1/2\,\beta_{24} \\ \beta_2 - 1/2\,\beta_{13} - 1/2\,\beta_{14} - 1/2\,\beta_{34} \\ \beta_3 - 1/2\,\beta_{12} - 1/2\,\beta_{24} \\ \beta_4 - 1/2\,\beta_{12} - 1/2\,\beta_{23} \\ \beta_{11} - 1/2\,\beta_{23} + 1/2\,\beta_{24} + \beta_{34} \\ \beta_{22} + 1/2\,\beta_{13} + 1/2\,\beta_{14} + 1/2\,\beta_{34} \\ \beta_{33} + 1/2\,\beta_{12} + \beta_{14} - 1/2\,\beta_{24} \\ \beta_{44} - 1/2\,\beta_{12} + \beta_{13} + 1/2\,\beta_{23} \end{bmatrix}$$

12.12. Suppose that there are four controllable variables and two noise variables. It is necessary to estimate the main effects and two-factor interactions of all of the controllable variables, the main effects of the noise variables, and the two-factor interactions between all controllable and noise factors. If all factors are at two levels, what is the minimum number of runs that can be used to estimate all of the model parameters using a combined array design? Use a *D*-optimal algorithm to find a design.

Twenty-one runs are required for the model, with five additional runs for lack of fit, and five as replicates for a total of 31 runs as follows. It should be noted that Design-Expert*'s D*-optimal algorithm might not create the same design if repeated.

Std	A	B	C	D	E	F
1	+1	+1	-1	+1	+1	+1
2	-1	+1	-1	+1	-1	-1
3	+1	-1	-1	+1	-1	-1
4	+1	+1	-1	-1	-1	+1
5	-1	+1	-1	-1	+1	+1
6	-1	+1	+1	+1	+1	+1
7	+1	+1	-1	-1	+1	-1
8	-1	-1	+1	+1	-1	-1
9	-1	+1	+1	-1	+1	-1
10	-1	+1	+1	-1	-1	+1
11	+1	-1	+1	+1	+1	+1
12	+1	+1	+1	+1	-1	+1
13	+1	-1	-1	-1	+1	+1
14	+1	+1	+1	-1	+1	+1
15	-1	-1	-1	-1	-1	-1
16	+1	+1	+1	+1	+1	-1
17	-1	-1	-1	+1	-1	+1
18	-1	-1	-1	+1	+1	-1
19	-1	-1	+1	-1	+1	+1
20	+1	-1	+1	-1	+1	-1
21	+1	-1	+1	-1	-1	+1
22	+1	+1	+1	-1	-1	-1
23	+1	-1	-1	-1	-1	-1
24	-1	+1	-1	-1	-1	-1
25	+1	+1	-1	-1	-1	-1
26	-1	-1	+1	-1	-1	-1
27	+1	+1	+1	+1	-1	+1
28	-1	-1	-1	+1	-1	+1
29	+1	+1	+1	+1	+1	-1
30	-1	-1	-1	+1	+1	-1
31	-1	+1	-1	-1	+1	+1

12.13. Suppose that there are four controllable variables and two noise variables. It is necessary to fit a complete quadratic model in the controllable variables, the main effects of the noise variables, and the two-factor interactions between all controllable and noise factors. Set up a combined array design for this by modifying a central composite design.

The following design is a half fraction central composite design with the axial points removed from the noise factors. The total number of runs is forty-eight, which includes eight center points.

Std	A	B	C	D	E	F
1	-1	-1	-1	-1	-1	-1
2	+1	-1	-1	-1	-1	+1
3	-1	+1	-1	-1	-1	+1
4	+1	+1	-1	-1	-1	-1
5	-1	-1	+1	-1	-1	+1
6	+1	-1	+1	-1	-1	-1
7	-1	+1	+1	-1	-1	-1
8	+1	+1	+1	-1	-1	+1
9	-1	-1	-1	+1	-1	+1
10	+1	-1	-1	+1	-1	-1
11	-1	+1	-1	+1	-1	-1
12	+1	+1	-1	+1	-1	+1
13	-1	-1	+1	+1	-1	-1
14	+1	-1	+1	+1	-1	+1
15	-1	+1	+1	+1	-1	+1
16	+1	+1	+1	+1	-1	-1
17	-1	-1	-1	-1	+1	+1
18	+1	-1	-1	-1	+1	-1
19	-1	+1	-1	-1	+1	-1
20	+1	+1	-1	-1	+1	+1
21	-1	-1	+1	-1	+1	-1
22	+1	-1	+1	-1	+1	+1
23	-1	+1	+1	-1	+1	+1
24	+1	+1	+1	-1	+1	-1
25	-1	-1	-1	+1	+1	-1
26	+1	-1	-1	+1	+1	+1
27	-1	+1	-1	+1	+1	+1
28	+1	+1	-1	+1	+1	-1
29	-1	-1	+1	+1	+1	+1
30	+1	-1	+1	+1	+1	-1
31	-1	+1	+1	+1	+1	-1
32	+1	+1	+1	+1	+1	+1
33	-2.378	0	0	0	0	0
34	+2.378	0	0	0	0	0
35	0	-2.378	0	0	0	0
36	0	+2.378	0	0	0	0
37	0	0	-2.378	0	0	0
38	0	0	+2.378	0	0	0
39	0	0	0	-2.378	0	0
40	0	0	0	+2.378	0	0
41	0	0	0	0	0	0
42	0	0	0	0	0	0
43	0	0	0	0	0	0
44	0	0	0	0	0	0
45	0	0	0	0	0	0
46	0	0	0	0	0	0
47	0	0	0	0	0	0
48	0	0	0	0	0	0

CHAPTER 13

Experiments with Random Factors

LEARNING OBJECTIVES

After completing this chapter, you will be able to:

1. Understand the difference between fixed and random factors.

2. Analyze data from a random effects experiment and estimate the components of variance.

3. Analyze data from a mixed model experiment and estimate the components of variance.

4. Generate the expected mean squares and determine appropriate test statistics for models with random factors.

KEY CONCEPTS AND IDEAS

1. Fixed factor

2. Random factor

3. Variance components

4. Expected mean squares

5. Mixed model

6. Restricted and unrestricted models

7. REML estimation of variance components

Exercises

13.1. An experiment was performed to investigate the capability of a measurement system. Ten parts were randomly selected, and two randomly selected operators measured each part three times. The tests were made in random order, and the data are shown in Table P13.1.

Table P13.1

Part Number	Operator 1 Measurements			Operator 2 Measurements		
	1	2	3	1	2	3
1	50	49	50	50	48	51
2	52	52	51	51	51	51
3	53	50	50	54	52	51
4	49	51	50	48	50	51
5	48	49	48	48	49	48
6	52	50	50	52	50	50
7	51	51	51	51	50	50
8	52	50	49	53	48	50
9	50	51	50	51	48	49
10	47	46	49	46	47	48

(a) Analyze the data from this experiment.

MINITAB Output

```
ANOVA: Measurement versus Part, Operator

Factor    Type     Levels   Values
Part      random     10      1    2    3    4    5    6    7
                             8    9   10
Operator  random      2      1    2

Analysis of Variance for Measurem

Source        DF        SS        MS       F       P
Part           9    99.017    11.002   18.28   0.000
Operator       1     0.417     0.417    0.69   0.427
Part*Operator  9     5.417     0.602    0.40   0.927
Error         40    60.000     1.500
Total         59   164.850

Source          Variance Error Expected Mean Square for Each Term
                component term (using restricted model)
  1 Part         1.73333    3   (4) + 3(3) + 6(1)
  2 Operator    -0.00617    3   (4) + 3(3) + 30(2)
  3 Part*Operator -0.29938  4   (4) + 3(3)
  4 Error        1.50000        (4)
```

(b) Find point estimates of the variance components using the analysis of variance method.

$$\hat{\sigma}^2 = MS_E \qquad \hat{\sigma}^2 = 1.5$$

$$\hat{\sigma}_{\tau\beta}^2 = \frac{MS_{AB} - MS_E}{n} \qquad \hat{\sigma}_{\tau\beta}^2 = \frac{0.6018519 - 1.5000000}{3} < 0 \text{, assume } \hat{\sigma}_{\tau\beta}^2 = 0$$

$$\hat{\sigma}_\beta^2 = \frac{MS_B - MS_{AB}}{an} \qquad \hat{\sigma}_\beta^2 = \frac{11.001852 - 0.6018519}{2(3)} = 1.7333$$

$$\hat{\sigma}_\tau^2 = \frac{MS_A - MS_{AB}}{bn} \qquad \hat{\sigma}_\tau^2 = \frac{0.416667 - 0.6018519}{10(3)} < 0, \text{ assume } \hat{\sigma}_\tau^2 = 0$$

All estimates agree with the MINITAB output.

13.2. An article by Hoof and Berman ("Statistical Analysis of Power Module Thermal Test Equipment Performance," *IEEE Transactions on Components, Hybrids, and Manufacturing Technology* Vol. 11, pp. 516-520, 1988) describes an experiment conducted to investigate the capability of measurements on thermal impedance (C°/W x 100) on a power module for an induction motor starter. There are 10 parts, three operators, and three replicates. The data are shown in Table P13.2.

Table P13.2

Part Number	Inspector 1			Inspector 2			Inspector 3		
	Test 1	Test 2	Test 3	Test 1	Test 2	Test 3	Test 1	Test 2	Test 3
1	37	38	37	41	41	40	41	42	41
2	42	41	43	42	42	42	43	42	43
3	30	31	31	31	31	31	29	30	28
4	42	43	42	43	43	43	42	42	42
5	28	30	29	29	30	29	31	29	29
6	42	42	43	45	45	45	44	46	45
7	25	26	27	28	28	30	29	27	27
8	40	40	40	43	42	42	43	43	41
9	25	25	25	27	29	28	26	26	26
10	35	34	34	35	35	34	35	34	35

(a) Analyze the data from this experiment, assuming both parts and operators are random effects.

MINITAB Output

```
ANOVA: Impedance versus Inspector, Part

Factor    Type    Levels   Values
Inspecto  random     3      1    2    3
Part      random    10      1    2    3    4    5    6    7
                            8    9   10


Analysis of Variance for Impedanc

Source         DF        SS         MS        F       P
Inspecto        2     39.27      19.63     7.28   0.005
Part            9   3935.96     437.33   162.27   0.000
Inspecto*Part  18     48.51       2.70     5.27   0.000
Error          60     30.67       0.51
Total          89   4054.40


Source         Variance Error Expected Mean Square for Each Term
               component term (using restricted model)
1 Inspecto       0.5646    3   (4) + 3(3) + 30(1)
2 Part          48.2926    3   (4) + 3(3) + 9(2)
3 Inspecto*Part  0.7280    4   (4) + 3(3)
```

```
4  Error         0.5111      (4)
```

(b) Estimate the variance components using the analysis of variance method.

$$\hat{\sigma}^2 = MS_E \qquad \hat{\sigma}^2 = 0.51$$

$$\hat{\sigma}^2_{\tau\beta} = \frac{MS_{AB} - MS_E}{n} \qquad \hat{\sigma}^2_{\tau\beta} = \frac{2.70 - 0.51}{3} = 0.73$$

$$\hat{\sigma}^2_{\beta} = \frac{MS_B - MS_{AB}}{an} \qquad \hat{\sigma}^2_{\beta} = \frac{437.33 - 2.70}{3(3)} = 48.29$$

$$\hat{\sigma}^2_{\tau} = \frac{MS_A - MS_{AB}}{bn} \qquad \hat{\sigma}^2_{\tau} = \frac{19.63 - 2.70}{10(3)} = 0.56$$

All estimates agree with the MINITAB output.

13.3. Reconsider the data in Problem 5.8. Suppose that both factors, machines and operators, are chosen at random.

(a) Analyze the data from this experiment.

Operator	Machine 1	2	3	4
1	109	110	108	110
	110	115	109	108
2	110	110	111	114
	112	111	109	112
3	116	112	114	120
	114	115	119	117

The following MINITAB output contains the analysis of variance and the variance component estimates:

MINITAB Output

```
ANOVA: Strength versus Operator, Machine

Factor    Type    Levels  Values
Operator  random    3     1    2    3
Machine   random    4     1    2    3    4

Analysis of Variance for Strength

Source            DF       SS       MS       F      P
Operator           2   160.333   80.167   10.77  0.010
Machine            3    12.458    4.153    0.56  0.662
Operator*Machine   6    44.667    7.444    1.96  0.151
Error             12    45.500    3.792
Total             23   262.958

Source          Variance Error Expected Mean Square for Each Term
                component term (using restricted model)
 1 Operator       9.0903    3   (4) + 2(3) + 8(1)
 2 Machine       -0.5486    3   (4) + 2(3) + 6(2)
```

```
3 Operator*Machine    1.8264    4    (4) + 2(3)
4 Error                3.7917         (4)
```

(b) Find point estimates of the variance components using the analysis of variance method.

$$\hat{\sigma}^2 = MS_E \qquad \hat{\sigma}^2 = 3.79167$$

$$\hat{\sigma}^2_{\tau\beta} = \frac{MS_{AB} - MS_E}{n} \qquad \hat{\sigma}^2_{\tau\beta} = \frac{7.44444 - 3.79167}{2} = 1.82639$$

$$\hat{\sigma}^2_{\beta} = \frac{MS_B - MS_{AB}}{an} \qquad \hat{\sigma}^2_{\beta} = \frac{4.15278 - 7.44444}{3(2)} < 0 \text{, assume } \hat{\sigma}^2_{\beta} = 0$$

$$\hat{\sigma}^2_{\tau} = \frac{MS_A - MS_{AB}}{bn} \qquad \hat{\sigma}^2_{\tau} = \frac{80.16667 - 7.44444}{4(2)} = 9.09028$$

These results agree with the MINITAB variance component analysis.

13.6. Reanalyze the measurement systems experiment in Problem 13.1, assuming that operators are a fixed factor. Estimate the appropriate model components.

The following analysis assumes a restricted model:

MINITAB Output

```
ANOVA: Measurement versus Part, Operator

Factor    Type    Levels    Values
Part      random   10        1     2     3     4     5     6     7
                             8     9    10
Operator  fixed     2        1     2

Analysis of Variance for Measurem

Source         DF        SS        MS       F       P
Part            9     99.017    11.002    7.33   0.000
Operator        1      0.417     0.417    0.69   0.427
Part*Operator   9      5.417     0.602    0.40   0.927
Error          40     60.000     1.500
Total          59    164.850

Source        Variance  Error  Expected Mean Square for Each Term
              component  term   (using restricted model)
1 Part          1.5836    4     (4) + 6(1)
2 Operator                3     (4) + 3(3) + 30Q[2]
3 Part*Operator -0.2994   4     (4) + 3(3)
4 Error         1.5000          (4)
```

$$\hat{\sigma}^2 = MS_E \qquad \hat{\sigma}^2 = 1.5000$$

$$\hat{\sigma}^2_{\tau\beta} = \frac{MS_{AB} - MS_E}{n} \qquad \hat{\sigma}^2_{\tau\beta} = \frac{0.60185 - 1.5000}{3} < 0 \text{ assume } \hat{\sigma}^2_{\tau\beta} = 0$$

$$\hat{\sigma}^2_{\tau} = \frac{MS_A - MS_E}{bn} \qquad \hat{\sigma}^2_{\tau} = \frac{11.00185 - 1.50000}{2(3)} = 1.58364$$

These results agree with the MINITAB output.

CHAPTER 14

Nested and Split-Plot Designs

LEARNING OBJECTIVES

After completing this chapter, you will be able to:

1. Design and analyze experiments with nested factors.

2. Design and analyze experiments with both nested and factorial (or crossed) factors.

3. Understand industrial situations where hard-to-change factors or other randomization restrictions lead to split-plot designs.

4. Design and analyze split-plot experiments.

KEY CONCEPTS AND IDEAS

1. Nested factor

2. Variance components

3. Two-stage nested design

4. General m-stage nested design

5. Expected mean squares

6. Whole plot

7. Whole plot error

8. Split plot (subplot)

9. Subplot error

Exercises

14.1. A rocket propellant manufacturer is studying the burning rate of propellant from three production processes. Four batches of propellant are randomly selected from the output of each process and three determinations of burning rate are made on each batch. The results follow. Analyze the data and draw conclusions.

Batch	Process 1				Process 2				Process 3			
	1	2	3	4	1	2	3	4	1	2	3	4
	25	19	15	15	19	23	18	35	14	35	38	25
	30	28	17	16	17	24	21	27	15	21	54	29
	26	20	14	13	14	21	17	25	20	24	50	33

MINITAB Output

```
ANOVA: Burn Rate versus Process, Batch

Factor            Type     Levels   Values
Process           fixed       3     1     2     3
Batch(Process)    random      4     1     2     3     4

Analysis of Variance for Burn Rat

Source          DF         SS          MS        F       P
Process          2     676.06      338.03     1.46   0.281
Batch(Process)   9    2077.58      230.84    12.20   0.000
Error           24     454.00       18.92
Total           35    3207.64

Source             Variance  Error  Expected Mean Square for Each Term
                   component  term  (using restricted model)
  1 Process                     2   (3) + 3(2) + 12Q[1]
  2 Batch(Process)    70.64     3   (3) + 3(2)
  3 Error             18.92         (3)
```

There is no significant effect on mean burning rate among the different processes; however, different batches from the same process have significantly different burning rates.

14.3. A manufacturing engineer is studying the dimensional variability of a particular component that is produced on three machines. Each machine has two spindles, and four components are randomly selected from each spindle. The results follow. Analyze the data, assuming that machines and spindles are fixed factors.

Spindle	Machine 1		Machine 2		Machine 3	
	1	2	1	2	1	2
	12	8	14	12	14	16
	9	9	15	10	10	15
	11	10	13	11	12	15
	12	8	14	13	11	14

MINITAB Output

```
ANOVA: Variability versus Machine, Spindle

Factor             Type    Levels  Values
Machine            fixed      3     1    2    3
Spindle(Machine)   fixed      2     1    2

Analysis of Variance for Variabil

Source              DF       SS       MS       F      P
Machine              2    55.750   27.875   18.93  0.000
Spindle(Machine)     3    43.750   14.583    9.91  0.000
Error               18    26.500    1.472
Total               23   126.000
```

There is a significant effect on dimensional variability due to the machine and spindle factors.

14.5. Consider the three-stage nested design shown in Figure 14.5 to investigate alloy hardness. Using the data that follow, analyze the design, assuming that alloy chemistry and heats are fixed factors and ingots are random. Use the restricted form of the mixed model.

	Alloy Chemistry											
		1						2				
Heats	1		2		3		1		2		3	
Ingots	1	2	1	2	1	2	1	2	1	2	1	2
	40	27	95	69	65	78	22	23	83	75	61	35
	63	30	67	47	54	45	10	39	62	64	77	42

MINITAB Output

```
ANOVA: Hardness versus Alloy, Heat, Ingot

Factor              Type    Levels  Values
Alloy               fixed      2     1    2
Heat(Alloy)         fixed      3     1    2    3
Ingot(Alloy Heat)   random     2     1    2

Analysis of Variance for Hardness

Source              DF       SS       MS      F      P
Alloy                1     315.4    315.4   0.85  0.392
Heat(Alloy)          4    6453.8   1613.5   4.35  0.055
Ingot(Alloy Heat)    6    2226.3    371.0   2.08  0.132
Error               12    2141.5    178.5
Total               23   11137.0

Source             Variance Error Expected Mean Square for Each Term
                   component term (using restricted model)
1 Alloy                         3  (4) + 2(3) + 12Q[1]
2 Heat(Alloy)                   3  (4) + 2(3) + 4Q[2]
3 Ingot(Alloy Heat)   96.29     4  (4) + 2(3)
4 Error              178.46        (4)
```

Alloy hardness differs significantly due to the different heats within each alloy. This is an important finding, for it implies that heat-to-heat variability is the primary cause of the variability in hardness.

14.13. A process engineer is testing the yield of a product manufactured on three machines. Each machine can be operated at two power settings. Furthermore, a machine has three stations on which the product is formed. An experiment is conducted in which each machine is tested at both power settings, and three observations on yield are taken from each station. The runs are made in random order, and the results are shown in Table P14.1. Analyze this experiment, assuming all three factors are fixed.

Table P14.1

Station	Machine 1			Machine 2			Machine 3		
	1	2	3	1	2	3	1	2	3
Power	34.1	33.7	36.2	31.1	33.1	32.8	32.9	33.8	33.6
Setting	30.3	34.9	36.8	33.5	34.7	35.1	33.0	33.4	32.8
1	31.6	35.0	37.1	34.0	33.9	34.3	33.1	32.8	31.7
Power	24.3	28.1	25.7	24.1	24.1	26.0	24.2	23.2	24.7
Setting	26.3	29.3	26.1	25.0	25.1	27.1	26.1	27.4	22.0
2	27.1	28.6	24.9	26.3	27.9	23.9	25.3	28.0	24.8

The linear model is $y_{ijkl} = \mu + \tau_i + \beta_j + (\tau\beta)_{ij} + \gamma_{k(j)} + (\tau\gamma)_{ik(j)} + \varepsilon_{(ijk)l}$

MINITAB Output

```
ANOVA: Yield versus Machine, Power, Station

Factor             Type    Levels   Values
Machine            fixed      3      1    2    3
Power              fixed      2      1    2
Station(Machine)   fixed      3      1    2    3

Analysis of Variance for Yield

Source                      DF        SS        MS        F       P
Machine                      2    21.436    10.718     6.25   0.005
Power                        1   845.698   845.698   492.96   0.000
Station(Machine)             6    33.583     5.597     3.26   0.012
Machine*Power                2     0.383     0.191     0.11   0.895
Power*Station(Machine)       6    29.208     4.868     2.84   0.023
Error                       36    61.760     1.716
Total                       53   992.068

Source                     Variance Error Expected Mean Square for Each Term
                           component term (using restricted model)
 1 Machine                           6    (6) + 18Q[1]
 2 Power                             6    (6) + 27Q[2]
 3 Station(Machine)                  6    (6) + 6Q[3]
 4 Machine*Power                     6    (6) + 9Q[4]
 5 Power*Station(Machine)            6    (6) + 3Q[5]
 6 Error                    1.716         (6)
```

14.19. Steel is normalized by heating above the critical temperature, soaking, and then air cooling. This process increases the strength of the steel, refines the grain, and homogenizes the structure. An experiment is performed to determine the effect of temperature and heat treatment time on the strength of normalized steel. Two temperatures and three times are selected. The experiment is performed by heating the oven to a randomly selected temperature and inserting three specimens. After 10 minutes one specimen is removed, after 20 minutes the second specimen is removed, and after 30 minutes the final specimen is removed. Then the temperature is changed to the other level and the process is repeated. Four shifts are required to collect the data, which are shown below. Analyze the data and draw conclusions, assuming both factors are fixed.

Shift	Time(minutes)	Temperature (F)	
		1500	1600
1	10	63	89
	20	54	91
	30	61	62
2	10	50	80
	20	52	72
	30	59	69
3	10	48	73
	20	74	81
	30	71	69
4	10	54	88
	20	48	92
	30	59	64

This is a split-plot design. Shifts correspond to blocks, temperature is the whole plot treatment, and time is the subtreatments (in the subplot or split-plot part of the design). The expected mean squares and analysis of variance are shown below. The following MINITAB Output has been modified to display the results of the split-plot analysis. MINITAB will calculate the sums of squares correctly, but the expected mean squares and the statistical tests are not, in general, correct. Notice that the Error term in the analysis of variance is actually the three factor interaction.

MINITAB Output

```
ANOVA: Strength versus Shift, Temperature, Time

Factor     Type    Levels   Values
Shift      random     4     1     2     3     4
Temperat   fixed      2     1500  1600
Time       fixed      3     10    20    30

Analysis of Variance for Strength
                                           Standard        Split Plot
Source            DF        SS        MS      F      P       F      P
Shift              3    145.46     48.49   1.19  0.390
Temperat           1   2340.38   2340.38  29.20  0.012   29.21  0.012
Shift*Temperat     3    240.46     80.15   1.97  0.220
Time               2    159.25     79.63   1.00  0.422    1.00  0.422
Shift*Time         6    478.42     79.74   1.96  0.217
Temperat*Time      2    795.25    397.63   9.76  0.013    9.76  0.013
Error              6    244.42     40.74
Total             23   4403.63

Source          Variance Error Expected Mean Square for Each Term
                component term (using restricted model)
1 Shift            1.292    7   (7) + 6(1)
2 Temperat                  3   (7) + 3(3) + 12Q[2]
3 Shift*Temperat  13.139    7   (7) + 3(3)
4 Time                      5   (7) + 2(5) + 8Q[4]
5 Shift*Time      19.500    7   (7) + 2(5)
6 Temperat*Time             7   (7) + 4Q[6]
7 Error           40.736        (7)
```

14.20. An experiment is designed to study pigment dispersion in paint. Four different mixes of a particular pigment are studied. The procedure consists of preparing a particular mix and then applying that mix to a panel by three application methods (brushing, spraying, and rolling). The response measured is the percentage reflectance of the pigment. Three days are required to run the experiment, and the data obtained follow. Analyze the data and draw conclusions, assuming that mixes and application methods are fixed.

Day	App Method	Mix 1	2	3	4
1	1	64.5	66.3	74.1	66.5
	2	68.3	69.5	73.8	70.0
	3	70.3	73.1	78.0	72.3
2	1	65.2	65.0	73.8	64.8
	2	69.2	70.3	74.5	68.3
	3	71.2	72.8	79.1	71.5
3	1	66.2	66.5	72.3	67.7
	2	69.0	69.0	75.4	68.6
	3	70.8	74.2	80.1	72.4

This is a split-plot design. Days correspond to blocks, mix is the whole plot treatment, and method is the subtreatment (in the subplot or split-plot part of the design). The following MINITAB Output has been modified to display the results of the split-plot analysis. MINITAB will calculate the sums of squares correctly, but the expected mean squares and the statistical tests are not, in general, correct. Notice that the Error term in the analysis of variance is actually the three factor interaction.

MINITAB Output

```
ANOVA: Reflectance versus Day, Mix, Method

Factor    Type     Levels   Values
Day       random      3     1    2    3
Mix       fixed       4     1    2    3    4
Method    fixed       3     1    2    3

Analysis of Variance for Reflecta
                                           Standard      Split Plot
Source         DF       SS        MS       F      P      F       P
Day             2     2.042     1.021    1.39   0.285
Mix             3   307.479   102.493  135.77   0. 000 135.75  0.000
Day*Mix         6     4.529     0.755    1.03   0.451
Method          2   222.095   111.047  226.24   0.000  226.16  0.000
Day*Method      4     1.963     0.491    0.67   0.625
Mix*Method      6    10.036     1.673    2.28   0.105   2.28   0.105
Error          12     8.786     0.732
Total          35   556.930

Source         Variance Error Expected Mean Square for Each Term
               component term (using restricted model)
1 Day            0.02406   7   (7) + 12(1)
2 Mix                      3   (7) + 3(3) + 9Q[2]
3 Day*Mix        0.00759   7   (7) + 3(3)
4 Method                   5   (7) + 4(5) + 12Q[4]
5 Day*Method    -0.06032   7   (7) + 4(5)
6 Mix*Method               7   (7) + 3Q[6]
7 Error          0.73213       (7)
```

CHAPTER 15

Other Design and Analysis Topics

LEARNING OBJECTIVES

After completing this chapter, you will be able to:

1. Use the Box-Cox method to select a transformation for the response variable in an experimental design.

2. Understand how the generalized linear model is an alternative way to analyze data where the normality and equality of variance assumptions are not satisfied.

3. Know how to analyze data from an unbalanced factorial design.

4. Use the analysis of covariance to account for an uncontrollable but measurable nuisance factor in a designed experiment.

KEY CONCEPTS AND IDEAS

1. Variance stabilizing transformation

2. Box-Cox method

3. Power family of transformations

4. Generalized linear model

5. Exponential family of distributions

6. Link function

7. Maximum likelihood estimation

8. Covariate

9. Analysis of covariance

10. Designing an experiment with covariates

Exercises

15.1. Reconsider the experiment in Problem 5.24. Use the Box-Cox procedure to determine if a transformation on the response is appropriate (or useful) in the analysis of the data from this experiment.

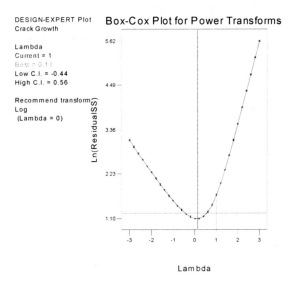

With the value of lambda near zero, and since the confidence interval does not include one, a natural log transformation would be appropriate.

15.2. In Example 6.3 we selected a log transformation for the drill advance rate response. Use the Box-Cox procedure to demonstrate that this is an appropriate data transformation.

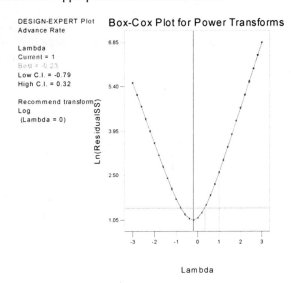

Because the value of lambda is very close to zero, and the confidence interval does not include one, the natural log was the correct transformation chosen for this analysis.

15.6. In the grill defects experiment described in Problem 8.30, a variation of the square root transformation was employed in the analysis of the data. Use the Box-Cox method to determine if this is the appropriate transformation.

The Box-Cox plot is shown below. Because the confidence interval for the minimum lambda does not include one, the decision to use a transformation is correct. Because the lambda point estimate is close to zero, the natural log transformation would be appropriate. This is a stronger transformation than the square root.

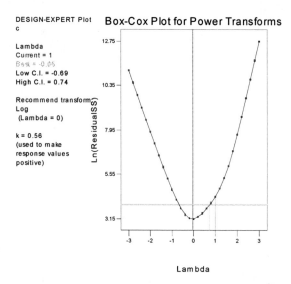

15.15. Four different formulations of an industrial glue are being tested. The tensile strength of the glue when it is applied to join parts is also related to the application thickness. Five observations on strength (*y*) in pounds and thickness (*x*) in 0.01 inches are obtained for each formulation. The data are shown in the following table. Analyze these data and draw appropriate conclusions.

Glue Formulation							
1	1	2	2	3	3	4	4
y	*x*	*y*	*x*	*y*	*x*	*y*	*x*
46.5	13	48.7	12	46.3	15	44.7	16
45.9	14	49.0	10	47.1	14	43.0	15
49.8	12	50.1	11	48.9	11	51.0	10
46.1	12	48.5	12	48.2	11	48.1	12
44.3	14	45.2	14	50.3	10	48.6	11

From the analysis performed in MINITAB, glue formulation does not have a statistically significant effect on strength. As expected, glue thickness does affect strength.

MINITAB Output

General Linear Model: Strength versus Glue

```
Factor    Type    Levels    Values
Glue      fixed      4      1 2 3 4

Analysis of Variance for Strength, using Adjusted SS for Tests

Source    DF    Seq SS     Adj SS     Adj MS      F       P
Thick      1    68.852     59.566     59.566    42.62   0.000
Glue       3     1.771      1.771      0.590     0.42   0.740
Error     15    20.962     20.962      1.397
Total     19    91.585


Term          Coef    SE Coef        T      P
Constant    60.089      1.944    30.91  0.000
Thick       -1.0099     0.1547   -6.53  0.000

Unusual Observations for Strength

Obs  Strength       Fit    SE Fit  Residual   St Resid
  3   49.8000   47.5299    0.5508    2.2701     2.17R

R denotes an observation with a large standardized residual.

Expected Mean Squares, using Adjusted SS

Source        Expected Mean Square for Each Term
 1 Thick      (3) + Q[1]
 2 Glue       (3) + Q[2]
 3 Error      (3)

Error Terms for Tests, using Adjusted SS

Source        Error DF  Error MS  Synthesis of Error MS
 1 Thick        15.00     1.397   (3)
 2 Glue         15.00     1.397   (3)

Variance Components, using Adjusted SS

Source     Estimated Value
Error             1.397
```

CPSIA information can be obtained at www.ICGtesting.com
Printed in the USA
BVOW02s1228310114

343521BV00022B/30/P